EAST ASIA
HISTORY, POLITICS, SOCIOLOGY, CULTURE

EDITED BY

EDWARD BEAUCHAMP

UNIVERSITY OF HAWAII

A ROUTLEDGE SERIES

East Asia
History, Politics, Sociology, Culture
Edward Beauchamp, *General Editor*

Modern Education, Textbooks, and the Image of the Nation
Politics and Modernization and Nationalism in Korean Education: 1880–1910
Yoonmi Lee

Problems of Democratization in China
Thomas G. Lum

The Unknown Cultural Revolution
Educational Reforms and Their Impact on China's Rural Development, 1966–1976
Dongping Han

Mao's Prey
The History of Chen Renbing, Liberal Intellectual
Jeanette Ford Fernandez

The Roots of Japan's Environmental Policies
Amy Wong

The Origins of the Bilateral Okinawa Problem
Okinawa in Postwar US–Japan Relations, 1945–1952
Robert D. Eldridge

South-South Transfer
A Study of Sino-African Exchanges
Sandra Gillespie

Studying the Enemy
Japan Hands in Republican China and Their Quest for National Identity
Christiane I. Reinhold

Self-Reliance or Self-Destruction?
Success and Failure of the Democratic People's Republic of Korea's Development Strategy of Self-Reliance "Juche"
Phillip H. Park

Third Delight
The Internationalization of Higher Education in China
Rui Yang

Struggle for National Survival
Eugenics in Sino-Japanese Contexts, 1896–1945
Yuehtsen Juliette Chung

Rethinking Japan's Identity and International Role
An Intercultural Perspective
Susanne Klien

Japan's Foreign Policy Maturation
A Quest for Normalcy
Kevin J. Cooney

Village, Market and Well-Being in a Rural Chinese Township
Tamara Perkins

History of Japanese Policies in Education Aid to Developing Countries, 1950s–1990s
The Role of the Subgovernmental Processes
Takao Kamibeppu

Engineering the State
The Huai River and Reconstruction in Nationalist China, 1927–1937
David A. Pietz

Struggle for National Survival

Eugenics in Sino-Japanese Contexts, 1896–1945

Yuehtsen Juliette Chung

Routledge
New York & London

Published in 2002 by
Routledge
29 West 35th Street
New York, NY 10001

Published in Great Britain by
Routledge
11 New Fetter Lane
London EC4P 4EE

Routledge is an imprint of the Taylor & Francis Group

Library of Congress Cataloging-in-Publication Data

Chung, Yuehtsen Juliette.
Struggle for national survival : Chinese eugenics in a transnational context, 1896–1945 / Yuehtsen Juliette Chung.
p. cm.
Inclues bibliographical references and index.
ISBN 0-415-93366-8
1. Eugenics—China—History. 2. Eugenics—Japan—History. I. Title.
HQ755.5.C6 C48 2002
363.9'2'0951—dc21 2002017782

To the memory of

Edward Tzuu Ch'ien

1940–1996

Contents

LIST OF TABLES

ACKNOWLEDGMENTS

EVERY TIME I READ A BOOK, I ALWAYS START WITH THE "ACKNOWLEDGMENTS," because I am curious about the work in its manufacturing process as well as in its finished form. Now it is my turn to recount the manufacturing course.

I am inexpressibly grateful for the opportunity and training I acquired in the Department of History, the University of Chicago. I owe a profound debt to my teachers for their generous and enthusiastic support. Prasenjit Duara opened my eyes and convinced me of the importance of the Japanese component in the understanding of modern Chinese history. Tetsuo Najita and Robert J. Richards have been a source of inspiration, and two chapters were written as a result of their stimulating questions. Leigh Van Valen from the Department of Ecology and Evolution walked with me patiently through the entire draft, and encouraged me in times when the idea of quitting seeped in.

For the research of this work, I benefited essentially from a ten-month intensive Japanese Language program fellowship from Stanford University at IUC, Yokohama, and Norma Field's support for it. Umemori Naoyuki of Waseda University and Takeshi Kimura of Yamaguchi University helped me access photocopies of entire runs of the journals *Yûseigaku* and *Yûsei undô*. Professors Hao Bin and Pan Naimu of Peking University assisted me in collecting materials from the University Library and the Beijing National Library. In the University of Chicago, I had John Crespi, Sarah Frederick and Katsuya Hirano share with me what they found in their own research trips to China and Japan. In addition, I had solid support from the East Asian Library and Inter-Library Loan service; special thanks go to Mr. Ma Tai-loi, Mr. Okuizumi Eizaburo, Mrs. Kuki and Ms Elaine Bao.

In the process of the painstaking writing, I was fortunate to have financial aid from the Committee on Chinese Studies and the Committee on Japanese Studies at the University of Chicago, and the Chiang Ching-kuo Foundation. During my stay in the Fairbank Center for East Asian Research at Harvard University as a Wang An Postdoctoral fellow in 1999-2000, I was given an opportunity to organize a workshop "Eugenics in East Asia: A Transnational Perspective (1868-1937)" sponsored by the Fairbank Center and the Reischauer Institute of Japanese Studies. I would like to express my gratitude to the faculty and staff of both institutes, and especially to Ruiko Conner for her warmth and friendship. I benefited fundamentally from the intellectual exchanges contributed by the workshop participants; special thanks go to Elizabeth Perry, Hue-Tam Ho Tai, Evelyn S. Rawski and Charlotte Ikels, Kenneth R. Robinson for their kindness and support, and to Sumiko Otsubo who has become ever since a great friend and a comrade in our pursuit of this difficult subject. I also incurred enormous debts to Philip A. Kuhn, Joshua Fogel, Joan Judge, Paul Cohen and Fan Fa-ti, who carefully read my dissertation and gave me many helpful comments and insights. Their constructive comments have been a vital source for current revision and further exploration. However, responsibility for views expressed in this book, as for errors, is mine alone. Chapter 5 will publish and appear in the collected volume entitled, *American Influence in Asia: Transpacific Relations in the Twentieth Century*, in Spring, 2002 by Greenwood Press. I thank Greenwood for allowing me to include this chapter, and Richard Jensen, Jon Davidann and Yone Sugita for their editorial comments.

I wish to express my heartfelt gratitude to my colleagues in the History Department at Hofstra for giving me more support than I perhaps deserved. I am very grateful to Farideh Kamali of Routledge and Edward Beauchamp for their confidence. Shelley Stephenson helped me with editorial comments, John Shea with production, and Alex Smiros with indexing in the preparation of manuscript. I thank them for their expertise and patience.

Without the moral support and friendship of Zhang Zhen, Magnus Fiskesjö, James St. André, Li Meng, Haiyan Lee, Chisato Hotta, Yasuko Sato, Bellette Lee, Sung Chia-Fu, Li Cho-ying, Rebecca Lien Chang, Wu Chyuan-yuan, Chang Yue-chin, Fu Dawie, Chen Kuan-Hsing, this work would not have been completed in such a swift pace. My family in Kaohsiung, Taiwan, and my husband Chien Chih-Yung are my anchorage from the unrest of the outside world. To them my deepest thanks for their faith in me.

This book is dedicated to the memory of Edward T. Ch'ien, my mentor and dear friend, who was a loving teacher and inspired my academic pursuit with his encouragement.

Struggle for National Survival

1

INTRODUCTION

EUGENICS—A GLOBAL INTEREST

The coinage of "eugenics" in 1883 by Francis Galton, the cousin of Charles Darwin, and the subsequent rise of the eugenics movement in the West, not only coincided with the preexisting intellectual milieu of Social Darwinism, but also corresponded to the greater discursive practice of Civilization that had existed since the middle of the nineteenth century. Together with Social Darwinism, the authority of Civilization, based upon Enlightenment and Christian values, determined that certain races, nations, ethnicities and classes were more civilized than others, and hence justified their domination over the lesser ones.[1] Eugenics in this context was considered as the science or technology of human betterment through the application of genetic laws designed so that the dominated could measure up to Civilizational standards, or the dominators to restrain the deplorable tendency of racial and national degeneration.[2]

According to the report on the International Commission of Eugenics published in 1924 in *Eugenical News*, there were fifteen full Commission members: Argentina, Belgium, Cuba, Czechoslovakia, Denmark, France, Germany, Great Britain, Italy, Netherlands, Norway, Russia, Sweden, Switzerland, and the United States. Additionally, there were seven other countries eligible for cooperation: Brazil, Canada, Colombia, Mexico, Venezuela, Australia and New Zealand.[3] If one observes scrupulously the global interest of eugenics, one finds nuances among the nations. However, it is impossible to exhaustively compare all eugenics movements in the world. In the following, I simply single out some of the European countries, the

United States and Russia as exemplars which some other countries, especially Japan and China, tried to emulate.

In the British Empire, the mental and physical deterioration perceived by eugenicists had to do with the effects of industrialization and urbanization that created or maintained an appalling cleavage between the upper and lower classes. According to his statistical analysis of British rural and industrial areas, Francis Galton warned that the descendants of contemporary townsmen would contribute only one fifth of the population of the country after two centuries. Thirty years later, the difficulty of recruitment for the Boer War reinvigorated, in the early twentieth century, the burning issues of physical degeneration and the decline of birth rate.[4]

Following the same track of industrialization, the United States also experienced a tremendous social change in the years around the turn of the century. The stretch of cities and slums, and the massive migrations from the countryside and from abroad, contributed to an ever-increasing perceived rate of prostitution, crime, alcoholism and mental defectiveness. As the white Protestants and eugenicists expressed their social and economic anxieties at the prospect of national deterioration, they always turned to the racial targets of blacks, Irish and Italian Roman Catholics and the Polish, Russian and German Jews, whom they thought to be hereditarily inferior. Their racist prejudice against blacks and Jewish and Catholic immigrants, as Daniel J. Kevles points out, figured more markedly in American than in British eugenics.

> British society was ethnically more-or-less homogeneous . . . and had not yet experienced the significant non-white immigration of later decades. . . . While British eugenicists talked of the threat of immigrants from Ireland and the Continent, they fretted a good deal more about the threat to the national fiber arising from the differential birth rate and the consequent weakening of their imperial competitive abilities in relation to France and Germany. British Eugenics was marked by a hostility decidedly more of class than of race.[5]

The problems of industrializing and urbanizing societies mentioned above in British and American settings also prevailed in France. However, in contrast to Britain and the United States, France was facing a more threatening issue, i.e. depopulation. Since the Franco-Prussian War of 1870–71, France had already lost military dominance on the Continent, and the German population was larger and growing faster than that of France. The military rivalry in Europe deepened the already pressing concern of depopulation, since mass conscription for military recruitment was getting more difficult than ever.[6]

The German version of eugenics favored the use of "race hygiene" (*Rassenhygiene*) as a means to strengthen racial stock, increase social productivity and seek cultural supremacy. Nonetheless, Sheila F. Weiss points out that before Hitler came to power in 1933, German eugenics was not primarily concerned with replenishing and improving the Nordic stock of Europe. Instead, *Rassenhygiene* was more preoccupied with class than with race. As opposed to other countries mentioned above, the concerns of German race hygiene were not fundamentally different from those of many other Western eugenics movements.[7] The perceived socio-political problems precipitated in Germany by industrialization included: 1) the rise of a radical labor movement threatening many middle- and upper-class Germans; 2) an increase in various types of criminal activity subverting the proper functioning of the state; 3) the growth of prostitution, suicides and alcoholism, and the increasing numbers of insane and feebleminded individuals, imposing a grave financial liability on the new Reich.[8]

One should not take eugenics as an exclusive rightist phenomenon and lose sight of the connection between eugenics and the left. Thus, it is important to insert here the Russian experience that would later have weight on the development of eugenics ideas in the world and in East Asia. As Loren R. Graham, Mark B. Adams and Diane B. Paul show us, the Soviet Union developed officially a eugenics movement in 1920. It became the first country in the world to ban eugenics in 1931 when Trofim Denisovich Lysenko rose to power and started a series of terrorizing campaigns persecuting Soviet geneticists. During this decade, Russia was a world leader in population genetics and agricultural genetics. Its eugenics movement was centered on the class concern because the Russian eugenicists had an explicit political agenda of opposing the Revolution and civil wars as dysgenic when these social upheavals drove waves of emigration of the aristocratic and intellectual classes.[9]

Socio-political problems that derived from industrialization and their dysgenic impact on populations were the combating targets which launched the Western eugenics movements. But both Japanese and Chinese intellectuals, in their initial introduction of eugenics ideas, tried to measure up to the Civilizational yardstick, since both Japan and China had still only gone halfway in their search for wealth and power.[10] When Fukuzawa Yukichi (1835–1901) brought up the topic of human improvement (*jinshu kairyô*) in his newspaper-essay collection *Fukuô hyakuwa* (One hundred essays of Mr. Fukuzawa) in 1896, he regarded the new advocacy of human application of animal breeding in the West, although as fantastical as aero-communication between Mars and the Earth, as a great inspiration for national strength, imperialist expansion and thus Civilizational progress. Fukuzawa considered the eugenic prospect of reproducing progeny, for instance, as moral as Sakyamuni, Confucius and Jesus Christ, or as brilliant as Isaac Newton, or as respectable as the military leader Katô Kiyomasa (1562–1611), was a far more sensible solution to racial improvement than the historical precedent approached by the Mongolian emperor Khubilai Khan (1215–1294) when he conquered China proper. It is said that Khubilai Khan originally planned to weed out entirely millions of the perceived inferior Han Chinese in the South and replace them with the better variety from the North. Khubilai Khan finally withdrew his plan by accepting the remonstration of his premier Yelü Chucai. Even though we do not have historical evidence of this story, we know for sure from James Lee's studies on the genealogy of the Qing imperial lineage that arranged marriage of royal elements, female infanticide and other controlled mortality measures, and utmost childcare of selected male infants formed an integrated mechanism of lineage selection.[11] In contrast, the modern eugenic technology of selecting desirable features – such as skin color, beauty, intelligence, robustness, and even the size of head and arms – was in Fukuzawa's eyes much more pleasant and feasible than the disheartening outlook of infanticide or of wiping out the entire population. Hence, it deserved more intellectual investment for civilizational progress (*bunmei shinpo*).[12]

The desire to enter the civilized world was felt even more acutely by Liang Qichao (1873–1929), whose intellectual weight in China was comparable to Fukuzawa's in Japan, and who was often acclaimed as the "Chinese Fukuzawa Yukichi." In addition to the lack of national consciousness and inability to participate in the nation-state community, Liang also held the savage practice of early marriage responsible for Chinese racial degeneration and their loss of dominance in the competitive evolutionary process, and hence keeping the Chinese from becoming fully civilized. Liang believed that the elimination of early marriage and family reform were the keys to halting the dysgenic cycle. As early marriage was used to demarcate the division between civilization and savagery, both China and Japan were almost on the same scale, falling behind the West and only slightly better than India. In a 1902 essay posted in the newspaper he himself ran, Liang tried to bring the issue of racial improvement and civilizational progress to the attention of other intellectuals and the literate public. Given the fact that Japan had made tremendous efforts in political reforms modeled after the Euro-American institutions and yet still fell behind in terms of civilizational level (*wenming chengdu*), Liang exhorted his compatriots not to take lightly the issue of early marriage as a vile communal custom and its link to the survival of the whole nation.[13]

With the threat of Western imperialism, Social Darwinism seemed to impose a national imperative for Japanese and Chinese intellectuals to import eugenics to East Asia as a cure for racial degeneration, and as a passport to enter the civilized world. However, we should not draw a conclusive picture that it was out of a nationalistic response to Western intrusion that eugenics in particular, and science in general, were imported and implanted in East Asia. As judged by the same criterion of modernization used to differentiate the scale of developed, developing and underdeveloped countries, the

Eastern experience of eugenics was regarded as either a perfect or an imperfect copy of a Western original. Such a view, however, is too parochial. Just as with different versions of modernity in the modern era, the emulation of the eugenics movement by developing and underdeveloped countries should be perceived as true diversity instead of degrees of imperfection, since there is no authority inherent in the original European version. The emulations, modifications, and new thoughts which jointly produce the different variations, are no less significant than the originals.

Comparativism—A Transnational Approach

In order to prevent upholding one single standardized view of eugenics which asserts its authority on the basis of a particular historical experience in Western countries, especially Britain and America, and to go beyond the conventional historical approach of value judgment as to the failure and success of eugenics, a comparative historical perspective in this study is not only desirable but also necessary. As Mark B. Adams points out, "When eugenics is studied in only one country, no matter how thoroughly, outcomes can appear to be overdetermined by a plethora of contextual religious, cultural, social, economic, institutional, and scientific variables that are difficult to rank in terms of their relative importance. When comparably sensitive case studies are set side by side, however, we may note broader patterns that would not be apparent from any single national case. This broader picture may, in turn, lead us to reevaluate and better understand eugenics in any particular country."[14] Instead of taking other partiesout of context merely in order to better one's self-understanding, this study's comparative historical perspective seeks to look at the two counterparts, China and Japan, in their own right. Yet it also focuses on their overlapping historical contexts, in which the development of eugenics in China and in Japan became interrelated during the first Sino-Japanese war (1894–95), which was a turning point for Sino-Japanese cultural exchange and technology transfer, and later converged in the historical event of the second Sino-Japanese War (1937–45), which was perceived by the eugenicists of both countries as a war of population.

My understanding of comparativism is a methodological approach to simultaneity which encompasses plural nonsynchronic consciousnesses and temporalities, in addition to the conventional use of comparativism as juxtaposition to locate parallels of similarity and difference. Inspired by Mikhail Bakhtin's dialogic conception of two contending tendencies, centrifugal forces and centripetal forces in any unitary language, and his idea of the polyphonic novel, which provides a picture of fluid interaction of plural consciousnesses of which none will be fully reduced as a mere object,[15] I perceive comparativism as a mode of writing to express the concept of simultaneity in order to locate intersubjectivity in both representations and reality, and better understand the conflictual development of Asian experiences of eugenics in particular and modernization in general.

Given that Japan and China as two parties (*aite*) for cooperation and competition in the same path of struggle for Civilizational progress in general, and for transplanting Western science in particular, the modern experience of Chinese eugenics can not be understood without taking into consideration the Japanese component as a source of inspiration, mediation and emulation. However, in order to avoid orientalizing the subject matter in this project, the comparative as well as transnational approach is designed to transcend the dichotomies of East versus West, representation versus reality, self versus other, etc. Instead of essentializing and oscillating between the dyadic parties, it tries to locate the simultaneity of power interplay in the Gramscian sense, which opens up the space of writing a history of ambiguity and difference. The trajectory of eugenics as I contour it in this work does not focus merely on individual agents and institutions, but also on the subjectivity amidst conflictural positions or even in the course of repetitious emulation.

Historically, the topic of eugenics was very popular between the 1880s and 1945, because its interest had been intertwined with the issues of race, gender, nationalism and scientific technology in the domains of national character, national body, and national survival. However, the subject was rarely discussed after World War II due to the stigmatization of the Nazi wartime experience. The recent English literature on Chinese eugenics includes: three books by Frank Dikötter, *The Discourse of Race in Modern China* (Stanford: Stanford Univ. Press, 1992), *Sex, Culture and Modernity in China* (Honolulu: University of Hawaii Press, 1995) and *Imperfect Conceptions: Medical Knowledge, Birth Defects and Eugenics in China* (New York: Columbia University Press, 1998); and Veronica Pearson, "Population Policy and Eugenics in China" in *British Journal of Psychiatry* (1995).[16] The particular interest of Dikötter's first book *The Discourse of Race in Modern China* is to see how the idea of race was articulated in Chinese eugenics. He includes a chapter on the Chinese eugenics movement, which is partly revised from his earlier article "Eugenics in Republican China."[17] In this chapter, he uses a narrative line of the inception, expansion and apogee of the Chinese eugenics movement. Such a narrative approach implies an assumption that the failure or success of the movement is judged by whether the eugenics bill ever became nationwide legislation. Such an assumption is based on the Anglo-American experience of the eugenics movement, although it is not necessarily a criterion by which American and British movements have been judged. Dikötter's second book *Sex, Culture and Modernity in China* focuses on the emergence of "sex" in Republican China as a topic of public discourse and its link to such social categories as "woman," "population" and "youth," within which modern biology, rather than the traditional Confucian cosmology, seems to serve as an epistemological underpinning at the time. Even though Dikötter dwells upon the theme of eugenics, he does not probe into the perplexing controversy among "Dr. Sex," Zhang Jingsheng, and the eugenicists Zhou Jianren and Pan Guangdan, which I do in the fifth chapter of this book.

As for the English literature on Japanese eugenics, Sumiko Otsubo's dissertation "Eugenics in Imperial Japan: Some Ironies of Modernity, 1883–1945" (Ohio State University, 1998) is the most thorough study thus far. It follows on the heels of pioneer Japanese monographs such as Suzuki Zenji's *Nihon no yûseigaku: Sono shisô to undô* (Eugenics in Japan: The trajectory of its ideas and movement, Tokyo: Sankyô, 1983), Matsubara Yôko's "Nihon ni okeru yûsei seisaku no keisei: Kokumin Yûsei-hô to Yûsei hogo-hô no seiritsu katei no kentô" (Dissertation, Ochanomizu Women's University, 1997) and Fujino Yutaka's *Nihon fashizumu to yûsei shisô* (Japanese fascism and eugenics thoughts, Kyoto: Kamogawa, 1998).[18]

These major studies contribute to our historical understanding of eugenics in the local contexts of China and Japan, and reveal an even more untrodden field of materials. Inspired by their historical insights, I organized a workshop "Eugenics in East Asia: A Transnational Perspective (1868–1937)" sponsored by the Fairbank Center for East Asian and the Reischauer Institute at Harvard University in May of 2000. The "transnational perspective" envisioned in this workshop extends beyond national boundaries so as to incorporate other possible forms of cultural interaction such as international coalitions, competition for scientific and technological advancements, and the emulation and circulation of eugenics knowledge. By disclosing flexible possibilities and their different combinations, we hope to recover source materials and writing left out of the trajectory of the nation-state's history.

The workshop participants Cheryl Barkey and Yuki Terazawa have finished dissertations on their respective topics of Chinese birth-control campaigns and Japanese reproductive medicine.[19] As for the other participants, Weili Ye is currently exploring the question of race encountered by Chinese overseas students in the United States in the 1920s, with a case study of Chinese eugenicist Pan Guangdan, while Sumiko Otsubo is expanding and placing Japanese eugenic interest in a much broader global context of migration. Jing Tsu of Harvard is working on her dissertation project on the literary representation of eugenics and science, and Kato Shigeo of Tokyo University is unraveling the paradox of individualism and totalitarianism conceived in the eugenics origin.

In comparison with the various studies mentioned above, my work is a historical investigation of different kinds of relationships between science and society, as seen through the eugenics movements in China and Japan from the 1890's to the 1940's, and of the legacies these movements bequeathed to the postwar and current developments of Chinese eugenics. I envision this project of eugenics as a specific case study against a greater background of the transmission of Western science. And I make a sort of nonstandard Lamarckian argument. By "Lamarckian," I mean the well known tenet of the "inheritance of acquired characters." Ideologically, the idea of Lamarck's time that somatic modifications acquired from an organism's development of particular habits would pass on to offspring under appropriate conditions, sustained a prospect of improving the human race through social reform and suggested that an improved environment would produce better people.[20] Despite the dynamics of eugenic ideas from the West, their reifications and historical ramifications were subject to their interaction with the culture of local habitats. Hence, the stigmatized attributes of eugenics in our current understanding should be modified, because each case was an individual outgrowth of its own historical context. Moreover, the comparative as well as transnational perspective focuses both on the contexts of these two interrelated movements in China and Japan, and on their specific local agendas. This particular approach seeks to locate the connections between the two movements in the form of linguistic and cultural influences and imperialism. It also explores the ways in which eugenics set forth themes and issues of local discourse, such as issues of coinage of scientific terminology, sexuality versus morality, population quality and growth, and preparation or deployment for war.

The style of each chapter in this eugenics study varies along with the content. This is simply because eugenics is an interdisciplinary subject and would be better approached by multidisciplinary writing. The book analyzes a number of aspects of the movement in depth. After the exploration of its global interest and its etymological process of name changes, chapter two then examines critically how organizational and professional structures, disciplinary agendas and interest groups interacted in the broader political and cultural contexts. Chapter three subsequently situates the debates on heredity and body in the arena of State control and social control of life. On the convergence of the birth-control movement with the eugenics movement, chapter four analyzes at the pedagogical and performative levels different perceptions of knowledge and reproduction among women. In the confrontation between Japanese and Chinese populations, chapter five lays out their different social constructions of race/ethnicity. Chapter six concludes with some lessons we learn from eugenics for the ongoing debates of bioethics and biotechnology, as well as an epilogue of research propositions to further explore the comeback of the Chinese eugenics discourse and campaigns in the 1980s and 1990s.

By analyzing simultaneously (in time) and yet nonsynchronically (in different paces of temporality) the eugenics movements in Japan and China, this study demonstrates on the one hand the dynamics of scientific ideas and practices across different spaces, and on the other hand the transforming power of local cultures. It also shows that eugenics is not a single unit of movement, and that it is possible to locate, from the so-called original Western version, a true diversity or difference out of a globalization process of ideas, scientific practice, and technology. This is because there is no authority inherent in the original version, and because the differences it generated were not diametrically antithetical to what it began with.

EUGENICS—LOCAL SIGNIFICATION

Eugenics (J. *yûsei,* C. *yousheng*) is striking for its manifold linguistic associations. The English meaning of "well-born" for eugenics, used for a book title by the historian of science Mark Adams, has its closest literal correspondence in the conventional Chinese and Japanese understanding of "prenatal care and teaching " (J. *taikyô,* C. *taijiao*) from well-off parents. However, the first time I ever encountered this term was in 1984–85 in Taiwan, when the government announced the national enactment and implementation of a Eugenics Bill in order to legalize abortion and improve population quality. Ever since then, almost every legal gynecological clinic has put out a sign of "*Yousheng*" on the streets to advertise as a suitable place to secure abortions. The visual impact of the signboard pressured me to detour via the center of the road whenever I had to pass these clinics. I am not sure if anyone else felt the same way as I did then, but eugenics was certainly equivalent to abortion in the sexual knowledge of my generation.

The picture became horrible in the late 1980s when the Taiwanese news media disseminated the image of a pregnant Chinese woman lying on an operating table and being forced by the PRC government to abort her six-month-old girl fetus. Later in Japan, where I was involved in my research project in 1994–5, my Japanese friends and language teachers asked about my topic. Their initial reaction was a pause of thirty seconds, after which they asked what is it about. After I told them that it dealt with racial improvement in the 1920s and 1930s till the end of the Second World War, silence pervaded for a while. Apparently, eugenics/*yûsei* was, as a concept, stigmatized by the Japanese wartime atrocities and their linkage to the Nazi holocaust.

When I went back to Taiwan for research, a friend devoted to the popular religion *Yiguan Dao* (the Unity sect of Five Teachings) gave me a puzzling lecture that the Buddhist doctrine of not killing (*bushasheng*) is the eugenic idea that I should focus on in my project. When I went to the National Central Library in Taipei to search for the subject "*Yousheng,*" I found a book entitled *Woguo zuiweida kexue zhi chuangzao faming* (The Greatest Scientific and Creative Invention in China). This book was published in 1985 and written by a self-named "professor" Liang Yinhe, who specialized in geomancy and was devoted to the divination theory of the *I Jing* (Book of Change). He advocated that these two great scientific heritages, geomancy and the *I Jing, should* be considered as eugenics (*yousheng kexue*) in order to produce better offspring and prevent cultural invasion from Western imperialism. Given the nationalistic overtone, one often finds a prevalent Chinese rationalization that any foreign novelties can be shown to specifically correspond to their Chinese equivalents or precedents.

As I probe deep into history I find names are not mere names. In post-war Japan, the name of eugenics was often associated with the practice of abortion and birth control counseling, in the teeth of a surging number of unwanted pregnancies resulting from prevalent adultery and interracial sexual affairs between American soldiers and Japanese women.[21] In China, eugenics did not re-emerge until the late 1970s and 1980s, when it coincided with the implementation of the "one-child" policy mandating that couples have no more than one child in urban areas (families in rural areas could have up to two children in very limited circumstances). Since then, eugenics has been tied closely with population control in order to achieve the goal of fewer but healthier babies with the prevention of birth and genetic defects.

Today, the eugenic ideas as embodied in the legal enactments are covered by the euphemistic emphasis of "motherhood protection" (J. *bosei hogo*, C. *muxing baohu*). This is the case both in Japan (with the 1947 Eugenics Protection Law, the 1966 Maternal and Infant Health Law, and the Protection Law of Maternal Body *"Botai hogohô"* in 1996, without any substantial change in contents and media coverage) and China (with the Maternal and Infant Health Law in 1995). From their discarding of the controversial English term "eugenics" so as to distance themselves from the Holocaust experience, and their defense of the Chinese characters *"Yousheng"* for their benign literal meaning of "healthy birth" or "superior birth," we can detect their abhorrence of uncovering the historical past of name changes. The Japanese ignore and unabashedly cover up the historical development of their own eugenics ideas, while its absence in historical textbooks has been favored by the official standing. The Japanese historiography of eugenics did not surface until the 1980s when Suzuki Zenji took the lead and published the first monograph of Japanese eugenics. Suzuki's interest in the eugenics study stems from the genealogical connection to his Marxist predecessors such as Yamamoto Senji and his cousin Yusuda Dokutaro in 1930s, and other Marxian historians of science such as Nakamura Teri, Tsukuba Hisaharu and Yonemoto Shôhe in 1970s.

In comparison, the Chinese are grossly ignorant of their historical legacy of a different eugenic signification and they are unable to mobilize the discursive resources in the process of policy-making for the cause of a better life that could be another literal meaning of *"Yousheng."* This section marks a first step to recover the abhorred past and explore cultural experience articulated specifically through an illustration of name changes for eugenic ideas in the linguistically interlocked context between China and Japan. The linguistic changes of naming do not come easily. The course of naming and renaming as negotiations and engagements in different battlefields of discourse was a politically charged power struggle.

Eugenic Ideas and Name Changes

In Japan, eugenic ideas were brought in by Fukuzawa Yukichi when in 1881 he referred to Francis Galton's *Hereditary Genius* in the *Jiji Shôgen* (Comment on Current Affairs). According to Suzuki Zenji's study, Fukuzawa appropriated Galton's views on hereditary human faculties in order to save the ex-samurai class from decline after the Meiji social reform that intermarriages between the upper and lower classes were officially allowed. Fukuzawa also made a couple of speeches on genetics and human improvement (*jinshu kairyô*) in 1896.[22] Fukuzawa's eugenics approach was further developed by his disciple Takahashi Yoshio, who proposed a theory of white and yellow interbreeding in an article, "Nihon jinshu kairyô ron" (The theory of improvement of the Japanese Race), in 1884. Takahashi's proposal of interbreeding was contested in 1886 by Katô Hiroyuki, who introduced and wrote about evolutionary theory. In 1905, the first issue of the journal *Jinsei* (Human Nature) published a Japanese translation of C. O. Whitman's article "The Heredity of Mental Power," in which Galton's ideas were further introduced. In 1910, Kôtoku Un'no's *Nihon jinshu kaizô ron* (The Improvement of the Japanese Race) appeared; it is regarded as the first comprehensive Japanese-language treatment of eugenics.[23] As eugenics ideas were imported collaterally with Social Darwinism, Japanese intellectuals appealed to them as a means of *fukoku kyôhei* (enriching one's country and strengthening military buildup) in a perilous time of international competition.

In China, prior to Yan Fu's translation of Huxley's *Evolution and Ethics* (*Tianyan lun*), where eugenics was mentioned as a means of improving racial stock though not for humans, Kang Youwei had around 1885–90 already advocated in the *Datong shu* (The Great Unity) that prenatal care and women's education on the one hand, and sterilization on the other hand, were the appropriate techniques for human selection.[24] Although the *Datong shu* was not published until 1935, Kang's eugenics proposal was recorded by his disciple Liang Qichao in *Kang Nanhai zhuan* (Biography of Kang Youwei), which was first published in 1906. In the *Datongshu*, Kang also advocated racial intermarriage in the hope of whitening the Chinese people. It is difficult to pinpoint the exact time when Kang acquired these ideas, and since he did not acknowledge his source of information, it is not easy to locate where he actually got his eugenics ideas. However, according to Kang's autobiography, he had been interested in Japanese reform and modernization since 1886. In 1896, after ten years of collection and reading, he compiled with his eldest daughter's help a bibliography of Japanese books (*Riben shumu zhi*), in which he included books on biological and social evolution and hygiene, and Katô Hiroyuki's speeches.[25] Moreover, several works of Katô Hiroyuki, and books on biology and medical studies with respect to prenatal care, women's hygiene, tuberculosis prevention, obstetrics and venereal disease were translated into Chinese before 1911.[26] Therefore, I suspect that Kang was not completely ignorant of Japanese eugenics knowledge.

In contrast to Kang Youwei's proposal, Liang Qichao suggested in 1902 a prohibition of early marriage, for five reasons: 1) it is harmful to health, because of an immature physical condition; 2) it is harmful to offspring because, according to statistics of the Western countries, the feebleminded and retarded, and criminals, were mostly the product of early marriages; 3) it is harmful in nurturing children because the parents are not experienced and educated enough to provide proper family discipline; 4) it is harmful to education because parents are unable to finish their studies; and 5) it is harmful to the national economy because younger parents need financial support from their own parents. What Liang was concerned with in this proposal was that the damage to people involved in early marriage, and its subsequent weakness in human resources, would devastate a fledgling national economy under the threat of imperialist encroachment.[27]

During the decades of the 1900s and 1910s, a variety of Japanese names for eugenic ideas, such as "*jinrui kairyô gaku*," "*jinshu kaizô ron*," "*minshû kaizen gaku*," or "*yûsei ron*" had been dispersed in various Japanese publications. In 1919 a new name, "*yûseigaku*" (literally, science of superior birth), emerged in Okaasa Jirô's *Saishin iden ron* (Modern genetics). It coexisted for a period with racial hygiene (*minzoku eisei*), but was accepted as the standardized Japanese term for eugenics when two major journals *Yûseigaku* (*Eugenics*) and *Yûseiundô* (*Eugenics Movement*) gained popularity. The term was subsequently exported to China.[28] Around the same time, a few terms like "*shanzhongxue*," "*shuzhongxue*," "*hunyin zhesixue*," and "*renzhong gailiangxue*," circulated in China. Among these names, "*jinrui kairyô gaku*" was the most stable term for eugenics, since it was included in the well-received *Dongwuxue dacidian* (Zoology dictionary), published in 1922 by the Commercial Press and edited by Du Yaquan (1873–1933, director of the science translation department in the Commercial Press and chief editor of the *Eastern Miscellany* and of a botany dictionary in 1917). According to the entry definition, "*jinrui kairyô gaku*" (eugenics) was defined as a branch of zoology which employed approaches of political control and social hygiene to investigate various causes and effects towards the goal of racial improvement.[29]

In both form and content, this dictionary definition was not very satisfying in the eyes of the Chinese eugenicist Pan Guangdang when he introduced the term "*yousheng*" for eugenics and established the Chinese eugenics association in 1924. Pan argued that the term "*yousheng*" used by the Japanese Eugenics Association and for its organizational journal was more appropriate than the above names, since it was phonetically and semantically closer to the original English word coined by Francis Galton in 1883. He also preferred this term over others such as "*renyixue*" (viriculture or hominiculture) and "*zhongyixue*" (stirpiculture), coined by Galton in 1873 and 1876 respectively, and over "*renzhong weisheng*" (Rassenhygiene), "*renzhong shengwuxue*" (Rassenbiologie), and "*jiazurenleixue*" (Familienanthropologie), popular in Germany and Scandinavia. Most importantly, Pan considered that use of this term could avoid the confusion which had arisen from the problematical word "J. *shu* or *tane*, or C. *zhong*" which had two implications, that is, the anthropological understanding of "race" and the biological sense of "seed" as "germplasm," such as "born with good seed (*danjiang jiazhong*)" in the *Book of Poetry* (*Shijing*). Pan argued that the anthropological category of "race" was not yet scientifically substantiated, and that eugenics as a newly established discipline, which only had a short history of fifty years, should not become entangled with the dubious claim of superior race. Every "color" of people fairly shared both good or bad germplasm distributed throughout its own population. For this reason, Pan concluded that eugenics should aim to track down the hereditary traits of human character, and their appropriate mechanisms for cultural selection, in order to uncover a better way of reproduction for human progress.[30]

Eugenics was a chaotic confusion of ideas developed earlier in the West, but Pan expected that since it was a relatively new subject in China, it could depart from such a path of development. To serve this purpose, Pan established in 1924 a demarcation of eugenics (*youshengxue*) specifically and exclusively to the extent that it looks like a manifesto: "It is not free love (*feiziyou lianai*); it is not sexual education (*feixing jiaoyu*); it is not public hygiene (*fei gongzhong weisheng*); it is not experimental marriage (*fei shiyan hunyin*); it is not a movement of anti-prostitution (*fei jinchang yundong*); it is not prenatal care and teaching (*fei taijiao*); it is not research on physical education (*fei tiyu yanjiu*); it is not compulsory marriage under the State's mandate; it does not advocate a massacre of the weak, the unruly and inferior (*bu tulu wanruo*); it does not aim to reproduce superman (*bu fanzhi chaoren*); it does not aim to reproduce genius (*bu shengchang tiancai*) to supply for the society's need; it does not aim to eliminate romanticism in love; it does not advocate animal breeding as a model for human rearing; it does not turn against any sensible concepts from sexual morality, marriage, love, family, reproduction and childrearing."[31]

In addition to this manifesto-like demarcation, Pan also identified the most eligible eugenic social elements, that is, the average and slightly above-average standard of intelligence, robust physique, moral decency and reproductivity, which often conformed to the upper-middle class. Although eugenics was not aimed at the reproduction of genius or uniform personality, it nonetheless targeted germplasm improvement in health, susceptibility to epidemic diseases, elimination of physical and temperamental defectiveness, creativity, resilience to external stimulation, sensational receptivity, intelligence and social coordination. Pan alerted his Chinese readers that eugenics should not be propagated as a magic bullet (*ruyidan*) for two reasons. First, eugenic studies were still in the preliminary stage of data collecting and theory testing, and had not yet reached any conclusive results. Second, the germplasm improvement was limited and valid only in the relative sense that it could be selected through the surviving quantity.[32]

The term eugenics/*youshengxue* was officially stabilized in the list of *Sociology Terminology* (*Shehuixue mingci*) as part of the project registering scientific terminology called *Kexue mingci huibian*. The list of *Sociology Terminology* was compiled by a group of Chinese sociologists, of whom Pan was one. It was sponsored by the National Translation Bureau beginning in 1937, and was finally promulgated by the Ministry of Education in 1941. However, despite all the efforts at stabilizing the name, the signifier "*yousheng*/*yûsei*" or eugenic rhetoric as it developed in China and Japan, derived manifold elusive signification. Ironically, some of these variants were diametrically opposed to what Pan had demarcated.

For instance, when the Japanese geneticist Tôyama Kametarô (1867–1918) was interbreeding French and Japanese silkworm cocoons in 1900 and published the result in 1906, it was Japan's first extended application of Mendel's laws from peas to animals. Based on his verification of applicability from plants to animals, Tôyama drew, one step further, an analogous relationship between animals and humans, and recommended that Mendelism be used for marriage criteria in order to remove undesirable human traits.[33] This animal-human analogous application was not rendered problematic by other Japanese geneticists and eugenicists.

When Margaret Sanger visited Japan and China in March and April, 1922, to promote birth-control campaigns, Chinese feminists and birth-control advocates embraced her proposal to free motherhood from slavery and make reproductive freedom and romantic love a foundation from which to build a new future for better offspring.[34] In the same fashion, Japanese feminists such as Yosano Akiko, Hiratsuka Raichô, Yamakawa Kikue and Ishimoto Shizue, despite the difference in their political ideologies, endorsed the eugenic pedagogy in exchange for support for birth-control campaigns. Their pedagogical participation in the eugenics movement won the Japanese women a solemn pledge in 1934: the National Eugenic Bill incorporated a stipulation to mandate a physical check-up before marriage and prevent marriage by people with venereal diseases, and it compelled the State to institute a health-care plan for motherhood protection.[35] Presumably, the feminist efforts to prevent venereal diseases would have been tied to the anti-prostitution movement, and thus cut across what Pan had demarcated.

In his 1923 article "*Shanzhongxue de lilun yu shishi*" (The theory and practice of eugenics), socialist eugenicist Zhou Jianren (1888–1984, Lu Xun's third younger brother, a Japanese Imperial University graduate and a popular translator of biology-and eugenics-related topics), urged that basic eugenic knowledge – such as the inheritance of genetic defectiveness and the danger of kin-marriage – be disseminated through sexual education before youths fell in love and yielded to tragic results. From Zhou's perspective, eugenics was not necessarily "not free love" as delimited by Pan Guangdan, but should rather be the guardian angel of securing freedom of love so as to provide good options for love.[36]

For Chinese educators like Zhang Junjun (1897–?, an ethnology and psychology professor and executive of Chinese Anti-tuberculosis Association in 1925) and Xu Shijin (1903–?, a lecturer at Beijing Medical College and worked later in the Hygiene Bureau in Shanghai in 1928), who together diagnosed dysgenic elements in the Chinese physical constitution, researches in physical education were an indispensable part of a comprehensive remedial prescription for national reconstruction (*minzu gaizao*). With a specific target on physical strength and physique, they recommended drastic dietary change, population control and environmental improvement, and physical training.[37]

This particular approach of emphasizing environmental factors was applied as a social-control project under the Lamarckian tenet of inheritance of acquired characters. It was espoused by both Chinese and Japanese hygienists, who saw the public-health problems as the primary factors endangering their nations and racial stock. They linked eugenics to population policies and made common cause with campaigns against alcoholism, opium addiction, tuberculosis, leprosy and venereal diseases as racial toxins. In the battle with leprosy in the 1920s and 1930s, the Japanese hygienists adhered to the eugenic rhetoric of "pure blood" and "racial purification" and finally gained the support of the Japanese eugenicists.

The union of hygienists and eugenicists coincided with the name change from the Japanese Eugenics Association (*Nihon yûseigaku kyôkai*) to the Japanese Racial Hygiene Association (*Nihon minzoku eisei gakkai*) in 1930. According to Nagai Hisomu (1876–1957), the director of the Japanese Racial Hygiene Association, one reason to change the name *yûseigaku* into *minzoku eisei* was because *yûseigaku* was overused by the regional eugenics movement advocated by Ikeda Shigenori (1892–1966). Nagai further differentiated that whereas eugenics focuses on genealogical and biometric investigation, racial hygiene includes not only the former approaches but also social sciences and social policies.[38] Moreover, both hygienists and eugenicists induced a rather aggressive proposal that the State should frame a eugenic law to prevent any proliferation and transmission of feeblemindedness, mental diseases, venereal diseases, alcoholism, tuberculosis and leprosy, and should mandate sterilization for these patients and sufferers before their marriage.

Converging effort from the Japanese eugenicists and hygienists on leprosy eradication invoked the name of "racial hygiene" in a rather literal sense, and rendered the same name used by the Chinese eugenicist Pan Guangdan into an intriguing signifier in his 1937 publication entitled *Minzu texing yu minzu weisheng*. I refrain from giving the English translation immediately after the Chinese characters, since what was intriguing was the specific meaning of the term "*minzu*" in Pan's conception. In contrast with the Japanese construction of the homogeneous and superior "Yamato race" Pan perceived that China was young as an ethnic-nation (*minzu*) because it had kept intermingling with other juvenile races throughout different historical periods.[39] Judging from the specific content of multi-ethnicity included in Pan's use of *zhonghua minzu*, and from his concrete non-Sinocentric proposals to reverse dysgenic selection – such as famine relief, elevation of economic productivity, impartial distribution of wealth, control of urbanization, family planning, and freedom of speech, press and political organizations[40] – we can be assured that Pan's version of *minzu weisheng* was different from the Japanese version of "racial hygiene," and closer to a rather inclusive grouping category. Hence, I feel more comfortable translating it as "national hygiene."

The divergent signification of eugenic ideas was even more acute in the confrontation of the two populations in the second Sino-Japanese War. Since both Japanese and Chinese eugenicists perceived the war as a war of population, they enacted eugenics bills as part of a national implementation of population policy. During this period, the Japanese Racial Hygiene ideology defended the Nazi's exclusion act against the Jews as "Rassenpflege (racial cultivation)," an effort of self-purification of their own racial blood, and not an aggressive plan to extinguish racially the Jews.[41] Pan Guangdan not only critiqued that particular version of racial ideology, but also abandoned the usage of the name *minzu weisheng*. In 1942, he modified the name for eugenics to "*minzu jiankang*" (national health) and promoted inter-ethnic mixture between the Han Chinese and the frontier minorities in order to rejuvenate the national stamina and improve the genetic makeup.

CONCLUDING REMARKS

In the context of the massive production of scientific terminology in modern Japan and China, Pan Guangdan's 1924 demarcating effort on eugenics as a signifier should be seen as a struggle of disciplinary identity for eugenics ideas. And Pan's shifts in career – from a zoologist in 1920s, to a sociologist in the 30s and 40s, and then to an ethnologist later in the 1950s and 1960s – say a lot more about the search for an institutional base for eugenics ideas than about the manifold elusive signification of "*yousheng/yûsei*." The recurrent loaning of Japanese terminology in the 1980's and 1990's shares unnecessarily a stigmatized past with Japan. That neither the Chinese nor the Japanese could claim a history of eugenic ideas of their own, also indicates an ineradicable component in the making of Sino-Japanese transcultural and translingual interaction from the Meiji ear. The loaning of terminology is reminiscent of Pan Guangdan's criticism of knowledge transplantation, such that the Chinese take Chinese ideographs for granted and the younger generation has no historical sense of the past as legacy.[42] Few contemporary Chinese eugenics activists recognize that Pan's non-Sinocentric eugenics proposals in 1930s continue to present convincing answers to current problems of Chinese wellbeing.

NOTES

[1] Prasenjit Duara, "The Discourse of Civilization and Pan-Asianism," in *Journal of World History*, (2001), vol. 12, no. 1, pp. 99–130.

[2] As Celeste M. Condit points out, Galton and other eugenicists employed the term *germ-plasm* rather than *gene* even after 1905 when the term genetics was coined, and the term gene was not coined until 1909. Even so, the term gene was not so visible in the public discourse of eugenics until the 1930s and 1940s. See Condit's *The Meanings of the Gene*, (Madison: University of Wisconsin Press, 1999), p. 6.

[3] Mark B. Adams ed. *The Wellborn Science: Eugenics in Germany, France, Brazil, and Russia* (Oxford: Oxford University Press, 1990), p. 5.

[4] See Richard A. Soloway, *Demography and Degeneration: Eugenics and the Declining Birthrate in Twentieth-Century Britain* (Chapel Hill: Univ. of North Carolina Press, 1990), pp. 38–47.

[5] Daniel J. Kevles, *In the Name of Eugenics:Genetics and the Uses of Human Heredity* (New York: Alfred A. Knopf, Inc., 1985), p. 76.

[6] William H. Schneider, "The Eugenics Movement in France" in *The Wellborn Science*, pp. 70–71.

[7] Sheila F. Weiss, "The Race Hygiene Movement in Germany, 1904–1945" in *The Wellborn Science*, p. 10.

[8] Ibid., p. 12.

[9] Loren R. Graham, "Science and Values: the Eugenics Movement in Germany and Russia in the 1920s" in the American Historical Review, vol. 82, no. 5, (1977), pp. 1133–1164. Mark B. Adams, "Eugenics in Russia, 1900–1940" in *The Wellborn Science*, pp. 153–216. Diane B. Paul, "Eugenics and the Left" in *The Politics of Heredity—Essays on Eugenics, Biomedicine, and the Nature-Nurture Debate*, (Albany: SUNY Press, 1998), pp. 11–35.

[10] See E. Sydney Crawcour, "Industrialization and Technological Change, 1885–1920" in *The Cambridge History of Japan* (Cambridge: Cambridge Univ. Press, 1988), vol. 6, pp. 385–450. According to E. Sydney Crawcour's study of Japanese economic development, even though it is true that the growth of modern sectors of industry and commerce accelerated in Japan after the Pacific War, these modern sectors were "still an infant nurtured by the traditional economy" from 1881 to 1913.

[11] James Lee and Wang Feng, "Nuptiality and Fertility among the Qing Nobility: Implications for Two Types of Preventive Checks" in *Qingdai huangzu renkou xingwei he shehui huanjing*, ed by James Lee and Guo Songyi, (Beijing: Beijing University Press, 1994), pp. 39–58.

[12] Fukuzawa Yukichi, in *Fukuzawa Yukichi shû* (Tokyo: Chikuma shobô, 1977), pp. 205–207.
[13] Liang Qichao, "Jin zaohun yi" together with "Xinmin yi" in *Yinbing shi heji* (Shanghai: Zhonghua shuju, 1932), Wenji, vol. 3, pp. 104–114
[14] Mark B. Adams, *The Wellborn Science*, pp. 6–7.
[15] See his "Discourse in the Novel" in *The Dialogic Imagination*, ed. by Michael Holquist and trans. by Caryl Emerson and Michael Holquist (Austin: University of Texas Press, 1981), pp. 259–422.
[16] Veronica Pearson, "Population Policy and Eugenics in China" was published in *British Journal of Psychiatry* (1995), 167, pp. 1–4. Dikötter latest book, *Imperfect Conceptions: Medical Knowledge, Birth Defects and Eugenics in China*, looks at the broader historical, cultural, social and political context of eugenics in China, but does not quite connect the broader context to the recent national enactment of a eugenics law in the People's Republic of China. Veronica Pearson, based on training in psychiatry, has an ongoing interest in eugenic ideas and their link to mental-health care and welfare in China, upon which I relies heavily.
[17] This article was published in *Republican China*, vol. 15, no. 1 (Nov. 1989).
[18] Sumiko Otsubo is currently teaching at Creighton University and her other publications include "Eugenics in Japan: Some Ironies of Modernity, 1883–1945," *Science in Context*, 11.3–4 (1998): 545–565 (co-author, James R. Bartholomew); "Feminist Maternal Eugenics in Wartime Japan," U.S.-Japan Women's Journal, English Supplement 17 (1999): 39–76, and "Engendering Eugenics: Feminists and Marriage Restriction Legislation in the 1920s," in Barbara Molony and Kathleen Uno, eds.,*Gendering Modern Japanese History* (Council on East Asia Studies, Harvard University, forthcoming). Her current research examines Japanese scientists and social reformers in the history of eugenic ideas, institutionalization, and legislation.
[19] Barkey's dissertation from UC Davis focuses on Chinese eugenics and its connection to the medical discourse and birth-control movement. Terazawa's dissertation is titled "Gender, Knowledge, and Power: Reproductive Medicine in Japan, 1690–1930" (University of California, Los Angeles, 2001).
[20] See more details in the introduction to the fourth chapter.
[21] See Sandra Buckly, "Body Politics: Abortion Law Reform" in *The Japanese Trajectory: Modernization and Beyond*, ed. by Gavan McCormack and Yoshio Sugimoto, (Cambridge: Cambridge University Press, 1988), pp. 205–217.
[22] These speeches were collected in *Fukuô hyakuwa* (One hundred essays of Mr. Fukuzawa), see Suzuki Zenji, "Geneticists and the Eugenics Movement in Japan," in *Japanese Studies in the History of Science*, (1975) vol. 14, p. 157.
[23] Ibid., p. 158.
[24] In Liang Qichao's *Kang Nanhai zhuan* (Biography of Kang Youwei), (Shanghai: Guangzhi shuju, 1906), pp. 31–32.
[25] See *Riben shumu zhi*, in the *Collection of Kang's Works* (*Kang Nanhai xiansheng yizhu huikan*), vol. 11, p. 62, p. 200, pp. 8–9, p. 558.
[26] See Tan Ruqian's bibliography, *Zhongguo yi riben zonghe mulu* (Hongkong: Zhongwen daxue, 1980).
[27] Wu Shenyuan, *Zhongguo renko sixiang shigao* (A history of Chinese population thought) (Zongqing: Xinhua shudian, 1986), pp. 269–71.

[28] Suzuki Zenji, *Nihon no yûseigaku—so no shisô to undô no kiseki* (Japanese eugenics: its ideas and locus of movement) (Tokyo: Sankyo, 1983), pp. 76–77. And for its export to China, see Gao Mingkai's *Hanyu wailaici cidian* (A dictionary of loan words) (Shanghai: Shanghai cishu, 1984), p. 394.
[29] Du Yaquan ed., *Dongwuxue dacidian* (Zoology dictionary), (Shanghai: the Commercial Press, 1922), p. 19.
[30] Pan Guangdan, "Yousheng gailun (General discussion on eugenics)," in Pan's *Collection (Pan Guangdan wenji)* (Beijing: Beijing daxue, 1993), vol. 1, pp. 250–265.
[31] Ibid.
[32] Ibid.
[33] Suzuki Zenji, *Nohon no yûseigaku*, pp. 67–68.
[34] See the special issue of the Lady's Journal (*Funü zazhi*), in June, 1922.
[35] See Suzuki Naoko ed., *Sengo bosei no yukue* (Tokyo: Domesu Shuppan, 1985) for the articles written respectively by Yosano Akiko, Hiratsuka Raichô, Yamakawa Kikue and Ishimoto Shizue.
[36] See Chen Changheng and Zhou Jianren, *Jinhualun yu shanzhongxue* (Evolution and Eugenics) (Shanghai: Commercial Press, 1923), pp. 63–78.
[37] Zhang Junjun, *Minzu suzhi zhi gaizao* (Chongqing: Commercial Press, 1944) and Chen Shijin, "Zhonghua minzu zhi jiankang" in *Kexue de minzu fuxing*, ed. by Zhu Kezhen et. al. (Shanghai: Zhongguo kexueshe, 1937), pp. 250–266.
[38] See Suzuki Zenji, *Nihon no yûseigaku*, pp. 150–151.
[39] Pan's judgment of China as a young ethnic-nation was based on Griffith Taylor's *Environment and Race: A Study of the Evolution, Migration, Settlement and Status of the Races of Man*, (London: Oxford University Press, 1927), pp. 187–195.
[40] See his work *Minzu texing yu minzu weisheng*, in his Collection, vol. 3, pp. 212–229.
[41] Furuya Yoshio, *Kokudo, jinkô, ketsueki*, (Tokyo: Asahi shimbunsha, 1941), pp. 176–177.
[42] Pan Guangdan, "Jinnian lai zhi zhishi jieshao" in *Dushu wenti*, originally was published in 1930 and later included in Pan's Collection, *Pan Guangdan wenji* (Beijing: Beijing Daxue, 1994), vol. 2, pp. 7–9.

2

Eugenics and Scientific Disciplines

In 1934, the prominent Chinese writer Lu Xun had been looking forward to the recent introduction of eugenics as a branch of scientific knowledge that could cure the sickly Chinese national character. However, viewing the semi-feudal China as a great vat of dye that altered any foreign elements, he was disappointed to find that eugenics, like every new science imported into China, had become an instrument used to secure the vested interests of the upper-class minority. He commented, "Physiology and anatomy were used to design devices of torture, and the knowledge of dynamics and electricity was employed to torture the Chinese revolutionaries. Geography was applied for geomancy, and chemistry was used for alchemy. Eugenics was used to legitimize the feudal system of arranged marriage between those of equal social status."[1]

I do not necessarily accept Lu Xun's value-oriented description of scientific development, but I do think that his comments indicate a general problematic phenomenon. That is, the globalization of science encounters social forces of local culture that frustrate the aspirations of modernizers around the world in their efforts to synchronize their "historical watches." These social forces have precursors, precedents, analogs and sources in the past that forced imported knowledge to adapt to local habitats.[2] This chapter analyzes these social forces and the ways in which eugenics was shaped and accommodated in various scientific disciplines.

After the initial stage of introducing eugenics ideas into East Asia, eugenicists tried to consolidate their ideas into scientific disciplines and in doing so were forced to encounter and engage in the local structure of science. In order to further local transformation under the conditions of limited resources and international competition, they favored the Galtonian biometric strategy over the Mendelian genetic application, to promote the former's presumable scientificity. Following the first chapter's discussion on Japanese influence in the coinage of eugenics, this chapter goes on to explore Japanese cultural impacts and imperialism shaping Chinese scientific disciplines in general and eugenics in particular. A comparative study of both science and eugenics in China and Japan aims to demonstrate the nonsynchronic nature of the globalization of science, and to serve as a prelude to their later confrontation in the second Sino-Japanese War, which I will discuss in Chapter five. In addition, such a comparative study is designed to tackle the continuity between local science and colonial science, like the parallel interface between nationalism and imperialism. As Prasenjit Duara has shown, this gives us historical insight into the ethnological production of frontier zones and territorial claims in the nationalist (or anti-colonial) agenda, which can be an expansionist act to coerce the peripheries and explore their resources in the names of national sovereignty, or in the Japanese case, East Asian Co-prosperity.[3]

THE AMERICAN CONNECTION AND NAZI CONNECTION

In 1975, the Japanese historian Suzuki Zenji conducted a pioneer comparative study on the American and Japanese eugenics development in 1900s-1920s.[4] In Japan, the eugenics-campaign activists were mostly physicians. In the formation of the Japanese Eugenics Society (*Nihon Yûsei Gakkai*) in 1925, there were 58 physicians, 4 biologists, 2 educators, 8 journalists and 8 in an unclassified category.[5] Unlike the United States, where nearly 50 percent of geneticists joined the eugenics movement in 1905–1915, Japanese geneticists, although advocating human improvement based on Mendelism at the time of its introduction, became less enthusiastic later. The main reason for geneticists to take an active part in the eugenics movement at its initial phase was to propagate and spread recognition of the need for genetic study. Compared with the volume of publications on animal and plant research, that on human genetics was relatively insignificant.[6] Suzuki gives three explanations for this: 1) there are not as many races in Japan as in the United States; 2) human genetics is technically difficult; 3) there is a high priority in Japan to promote staple crops such as the rice plant, and industrial organisms, especially the silkworm.[7]

Suzuki's comparativism raised a question of "Why America and Japan? Why not Germany and Japan?" Moreover, Suzuki's first two explanations of the trend of Japanese genetics cannot apply in the Chinese case. First, there are and have been various ethnicities in China. The Chinese Eugenics Association (*Zhongguo yousheng xuehui*) was founded in 1924 by Pan Guangdan (1898–1967), a biologist and a student of the American eugenicist Charles B. Davenport at Columbia University. The Chinese Eugenics Association, as it is mirrored in the short-lived journal *Yousheng yuekan*

(Eugenics monthly) from June, 1931, to March, 1932, was rather a one-man operation: a fact confirmed by Pan Naimu, Pan's daughter and compiler of his collected works, in my interview with her in Beijing, September, 1997. According to her, Pan was once a Christian and was funded by the YMCA to set up the eugenics journal. Later, he changed his belief and in April 1932 started another eugenics journal named *Huanian* (an abbreviation from the phrase helping the Chinese nation to reach maturity). Other intellectuals such as political activists, sociologists and scientists contributed to the journal in the form of social debate and discourse rather than institutional membership. In addition to Pan's own journals, eugenic interests were also expressed by physicians in the major journal of the Chinese Medical Association and by cultural theorists in the journal *Wenhua jianshe* (Cultural Construction) which I will discuss in Chapter three.

Second, the quality of Chinese genetics education and research was high, because Chinese geneticists were in the mainstream of genetics as it moved from the classical Morgan school to population genetics and the evolutionary synthesis.[8] However, Chinese geneticists were hardly involved at all with any campaign of racial improvement. According to Laurence A. Schneider's studies, Chinese genetic studies during the Republican period (1912–49) were simply centered on plants and animals.[9] Intriguingly, Chinese genetics received financial support from the Rockefeller Foundation and the China Foundation for the Promotion of Education and Culture, an organization which arose out of the USA's second remission of the Boxer Indemnity. If the Chinese genetic establishment borrowed from the American scientific model, which was shaped especially under the auspices of the Rockefeller Foundation (the RF "almost single-handedly paid for the establishment of modern biology and medicine in the United States before World War II,"[10]), why was it that Chinese geneticists, as recipients of the RF grants, did not follow their American geneticist counterparts, as many as half of whom joined the eugenics campaign?[11]

Based on previous scholarly production, it is easily assumed that only Japan and Nazi Germany constituted suitable parallels for a comparative study of eugenics in fascist regimes. This is an important and valid question for further exploration, but it does not necessarily attest to historical fact from a simultaneous transnational perspective. New studies of eugenics literature provide interesting evidence of American cooperation with Nazi Germany. Robert N. Proctor reveals that America was the first to turn eugenics ideas into reality, when it legalized sterilization of the mentally ill and criminally insane in Indiana as early as 1907. German racial hygienists had ever since looked upon the American and British achievements as the models for emulation. America's 1924 Immigration Restriction Laws, sterilization program and the anti-miscegenation laws in many states of the American South not only impressed German racial scientists, but also became an encouraging precedent for them to justify their eugenics campaigns. As a latecomer from 1933, the Nazis administered radical racial measures in its attempts to purify the German race of its Semitic elements, which in turn impressed American eugenicists such as Harry H. Laughlin and Charles Davenport, and aroused their envy for Nazi efficiency.[12]

Stefan Kuhl likewise documents that as American eugenicists admired the rapid development of the Nazi eugenics program, a number of prominent figures went to Germany for firsthand information, in order to prepare for any possible criticism they would receive should they implement similar measures. When these eugenics leaders—such as William W. Peter, Marie E. Kopp, Clarence Campbell, Charles M. Goethe, Marion S. Norton, Lothrop Stoddard, and the geneticist T. U. H. Ellinger—returned to the United States, they became the most ardent defenders of the Nazi racial purification campaigns. Stoddard and Ellinger, in particular, through their personal connection with the upper levels of Nazi authority, almost became the American mouthpiece of the German race policies during the wartime period.[13] Even though there was another group of American reform eugenicists who perceived themselves to be rather scientific-minded and hence distant from the mainline propaganda of ethnic selection, their bottom-line support for the Nazi sterilization program and their racist mindset functioned to artificially separate their science from mainline politics, and thus helped stabilize the Nazi regime.[14]

In the course of development of World War Two, American eugenicists were surprised at the Nazi German aggression against nations of the same "white Nordic stock." The break between German and American eugenicists only occurred after the United States entered the War in the Pacific following the Japanese attack at Pearl Harbor on December 7, 1941. While German racial hygienists were preoccupied with extensive extermination programs, the American Eugenics Society was barely keeping alive the organ publication *Eugenic News* around this period. Immediately after the War, German-American contact resumed and German eugenicists asked for American support to resume their academic positions and to return to the International community of science. In the 1946 Nuremberg Trial, Karl Brandt, who was responsible for the Nazi sterilization programs and for the elimination of thousands of "lives not worth living," defended himself by enumerating the American precedents of eugenics measures. The American prosecutors at this trial were not entirely unsympathetic to these arguments.[15]

The major driving force linking the American and Nazi connections was the extensive financial support of American foundations, the most important of which was the Rockefeller Foundation in New York. Even before World War One, the Rockefeller Foundation was interested in eugenics, and had ventured programmatic sponsorship in the areas of medicine, biochemistry, biophysics, and psychology in the United States. Rockefeller monies also went to

European countries, such as Britain, Demark, Norway, and especially Germany. Starting from 1926, the Rockefeller Foundation supported German eugenicists and until 1940 played an essential role in sponsoring major eugenic institutes, such as the Kaiser Wilhelm Institute for Psychiatry in Munich and the Kaiser Wilhelm Institute for Anthropology, Eugenics, and Human Heredity.[16] With Rockefeller aid, Germany and the United States had surpassed Great Britain as the leading forces of the international eugenics movement by 1930.

Compared to the Rockefeller patronage for research in racial hygiene and human genetics in Germany and other European countries, the grants to Japan and China for similar research were almost non-existent. Although the Rockefeller Foundation had made a total contribution of $44,944,665 to China from 1915 to 1947, the money mainly went towards establishing and maintaining the Peking Union Medical College, and financing national public health programs and the Chinese Mass Education movement.[17] The trifling remainder went mostly towards support for genetic studies on plants and animals.[18]

In Japan, between the 1890s and 1925, genetics research was largely agriculturally oriented, for instance, on rice, silkworms, wheat and Drosophila. Institutionally, this research was mostly produced from the Tokyo Imperial Agriculture University (1890) and Tôhoku Imperial Agricultural University (1911), where the improvement of staple crops was the priority.[19] As eugenics became a popular cause for social movement from the mid-1920s to the 1940s, Japanese eugenics leaders raised funds for their campaigns and research activities through personal channels and membership income. According to Sumiko Otsubo's study, Gotô Ryûkuchi (1887–), who founded the journal *Yûseigaku* (Eugenics), received support from American eugenicists such as Charles Davenport and Harry H. Laughlin for campaign advice and scholarly exchange to promote eugenics knowledge. With Davenport and Laughlin's permission and publication exchange program, the journal *Yûseigaku* was able to improve its publication quality and subscription revenue by adding a column of correspondence between the Japanese editor and the American advisors. It also distributed the Japanese translation of current news and articles from the Eugenics Record Office's *Eugenical News* and other pamphlets periodically sent by Davenport and his colleagues at Cold Spring Harbor, New York.[20]

Without much direct monetary assistance, American influence on eugenics in China and Japan was exercised through networking and a mentor-student metaphor. The success of the American eugenics movement in marriage restriction and sterilization legislation served as a source of inspiration for civilizational progress and provided a great model that Chinese and Japanese eugenicists were longing to emulate.[21]

The transnational convergence of eugenics between China and Japan was diverted later in the confrontation of the second Sino-Japanese war. Just as they were surprised at German aggression, American eugenicists would not understand Japanese aggression against the nation of the same "Yellow race." While Japanese people were becoming the enemy, the American eugenics movement was on the verge of a struggle for survival due to the bad reputation acquired from the Nazi connection. As Laughlin died in 1943 and Davenport in 1944, American eugenicists in the wartime period were unable to provide a better solution to a "war without mercy."

Given the above global context of American and Nazi connections, Suzuki's judgment of "technical difficulty" acquired a concrete meaning of limited resources in human genetics studies in the Japanese and Chinese eugenics context. The following sections provide a closer look at the interaction between eugenicists and geneticists within the disciplinary structure of science.

TRANSPLANTATION: MEDICINE AND BIOLOGY

In China, despite the presence of a flourishing traditional culture of medicine and biology, the modern disciplines of Medicine and Biology were not a natural outgrowth from the native soil. In the early 1910's, there were only 45 to 55 Western/Japanese-trained Chinese medical professionals; the rest of the 450 or so practitioners were Western medical missionaries and physicians in private practice. However, by 1920 the total number of Western-medicine physicians surged to 1500, out of whom about 900 were Chinese.[22] The key mechanism propelling this growth was the missionary establishment of 23 medical colleges and 36 nursing schools, pharmacological academies and midwifery schools between 1900 and 1915.[23]

Of primary importance was the founding of the Peking Union Medical College in 1915, when the Rockefeller Foundation took over from the previous joint management of the London Missionary Society, the American Presbyterian Mission and the American Board of Mission, the Board of Foreign Missions of the Methodist Episcopal Church and the Society for the Propagation of the Gospel, and the Medical Missionary Association of London. Once the Rockefeller Foundation stepped into science education in China, it changed the landscape of the predominantly missionary monopoly. This happened even though the Foundation was initially inspired and encouraged by the missionaries to help improve their evangelizing enterprise in China in the face of recent Chinese xenophobic reactions such as the Boxer Rebellion, the boycott of American goods in protest over immigration restrictions, and Chinese students preferring to undertake their overseas education in Japan rather than in America.[24] The Foundation's interest in sharing the development of China's resources coincided with an expanding American econo-political involvement in China, an anxiety over the disappearing

American frontier, the religious zeal of Christianity's perceived responsibility to the "heathen" world, and an optimistic reform design of changing the world by American progressive ideology.[25]

With a faculty formed from a mixture of missionary physicians and Chinese students trained overseas, the establishment of the Peking Union Medical College (hereinafter PUMC) figuratively marked a watershed in Chinese professionalization of modern medicine. It also provided a mechanism to produce and reproduce its own practitioners, to the extent that it transformed the social relation of physicians in particular and scientists in general with society as a whole. Although later in the 1930's it yielded to local demand for applied science through vocational training of physicians and public-health medicine, PUMC's initial phase as an institution was meant to be a transplantation of the "Johns Hopkins" institutional model, which was focused on integrating clinical practice with laboratory research. Permeably infused with German influence, the Johns Hopkins University Medical School raised entrance standards, emphasized laboratory and demonstration techniques, and had a faculty with a full-time commitment to the university. These reforming approaches led to a requirement of a four-year liberal-arts college education before admittance to the medical school at Johns Hopkins; this requirement became a paradigmatic practice for American medical education in a series of reforms in the 1910's.[26]

Hence, in order to transpose this "Johns Hopkins" high-quality model into the less viable Chinese context, the Rockefeller Foundation had, in addition to costly expenditures on laboratory equipment and building facilities, to cope with several difficulties. First, even though the Foundation emphasized the priority of upholding physicians' professional standards over religious devotion, it still had to show its willingness to cooperate with the missionaries without compromising its own scientific approach. Second, quite distinct from the missionary schools' indiscriminate admission policy, the Foundation's PUMC and its administrative organ China Medical Board focused on attracting the Chinese elite who were able to pay tuition twice as high as that of other prestigious colleges such as Qinghua University and who had fulfilled four years of college study of biology, physics and chemistry before admission.[27] Therefore, to overcome the above problems, the Foundation placed PUMC in a leading role in the missionary medical network and used it to retrain missionary physicians and upgrade them with the latest laboratory and research developments.

Furthermore, the Foundation obliged the missionary doctors to convert their home institutions into adequate preparatory schools in order to provide PUMC their best Chinese medical students. In return, the Foundation provided grants from 1915 to 1951 to some of the most outstanding missionary universities, such as Yenching University, the University of Nanking, Chinling College, Xiangya Medical School (Yale-in-China), Shantung Christian

University and Lingnan University. Even though the Foundation was not able to set up a unified national teaching standard among the various types of medical schools (for instance, the Japanese, English, and Continental models, as well as indigenous schools), it managed to use PUMC as a means of restructuring the production network of medical science.[28] After this point in time, Chinese students no longer had to seek overseas science education as the only means of obtaining high-tech training at the doctorate level.[29]

Without confining itself in the medical domain, the Rockefeller China Medical Board determined to go beyond its missionary network and safeguard a steadier flow of incoming competent students by advancing the quality of science education throughout universities and colleges in China. The China Medical Board's initial shift in the early 1920's was to fund science departments of public and private Chinese universities like Nankai University in northeast China, and National Southeastern University in Nanjing. The China Medical Board administered funds for physical facilities such as science buildings and research and teaching equipment and, in addition, provided fellowships for purposes such as bringing in American scientists and educators to PUMC for course work, and for regularly reviewing the most recent research for scientists within the network. As Laurence A. Schneider points out, later in the decade of 1925–1935, "the policy trends of the China Medical Board shifted from funding foreign institutions in China toward Chinese institutions; and away from medicine and hospital plants to basic science education and bio-medical research."[30] In this decade, as Chiang Yung-chen's most-up-to-date study shows us, the Rockefeller Foundation had shifted its focus of support toward social engineering as a global extension of Western social science movement.

The Rockefeller China Medical Board would not have been able to complete its role as a driving force for science-education networking, without cooperation from the China Foundation for the Promotion of Education and Culture, which was established in 1924 to administer twelve million dollars of America's second remission of the Boxer Indemnity. With the effective liaison of Roger Greene and Paul Monroe, who served in both organizations, the Rockefeller China Medical Board's science policies and programs were often endorsed and accommodated by the China Foundation.[31] Nonetheless, instead of targeting specifically the preparation of students to meet the admission requirements for medical schools, the China Foundation aimed to advance general (from colleges to secondary) science education and research and to improve library facilities.

According to the 1935 report of Ren Hongjun (1886–1961), a renowned scientist who served in both the China Foundation and the Science Society of China, in the decade 1925–35 the China Foundation not only endowed science-chair professorships and financed their research projects in geology, archaeology, zoology, botany and chemistry, but also provided fellowships for scientists to conduct research all over the world (see tables 1 and 2).[32] In addition, the Foundation accomplished five important tasks: 1) consolidating and restructuring the old Metropolitan Library into the National Library of Beijing; 2) fully subsidizing the establishment of Jingsheng Biological Research Laboratory in Beijing, run by the Science Society of China, and supporting their other biological institute in Nanjing and Sun Yat-sen's University Botanical Research Institute in Guangdong; 3) setting up a Sociological Research Institute on the basis of seed money donated by an American source; 4) translating natural science textbooks and later social science and humanity texts in order to stabilize and standardize translations of various scientific terminology; 5) supplying science instruments for high schools to consolidate their teaching facilities.[33]

Table 1 . Individual research projects sponsored by China Foundation

Year/ subjects	1928	1929	1930	1931	1932	1933	1934	Total
astronomy/ meteorology/ geography	2	3	5	4	3	4	6	28
mathematics/ physics/ chemistry	7	11	15	19	19	19	20	110
biology	11	21	26	20	22	23	23	146
Total	20	35	46	43	44	46	49	283

Table 2 . The locations of research projects of Table 1

Year/ country	1928	1929	1930	1931	1932	1933	1934	Total
China	1	20	18	18	18	16	17	118
U.S.A.	6	7	11	14	14	14	11	77
Britain	1	1	3	1	1	4	4	15
France	2	5	6	6	4	4	8	35
Germany	0	2	7	3	6	5	2	25
Other European countries	0	0	1	1	1	3	7	13
Total	20	35	46	43	44	46	49	283

From the two tables, we can see that biology seemed to be predominantly emphasized by the China Foundation as an area of science investment. Like geology and paleontology (not in the tables), much of biology is grounded within local fauna and flora; therefore, it is easily targeted as a particularly efficient local science that can compete in the international scientific arena within a short period of time.[34] With the private financial aid from the Foundation, the Science Society's Biological Research Laboratory and Institute stood out from the biology departments in eleven universities and was transformed into a cradle in which second-generation biologists were nurtured locally. Subsequently, these biologists assumed positions in the biology faculty in the universities and other academic research fellowships in the research institutes throughout the nation.[35]

The science infrastructure constructed by the two pillars of the Rockefeller Foundation and the China Foundation would not have functioned effectively had the Science Society of China not been there for community networking and participating in the process of policy-making.[36] Metaphorically, no one foot of the scientific "tripod" could dominate the other two. The Science Society of China was founded in 1914 in Ithaca, New York, initially by nine Chinese students in Cornell University, who later achieved prominence in various professions: Ren Hungjun, Hu Shi, Zhao Yuanren, Hu Mingfu, Bing Zhi, Zhou Ren, Yang Quan, Guo Tanxian and Jin Pangzheng.[37]

Before the headquarters of the Society moved back to Nanjing, China, in 1918, the membership of the Society was predominantly formed by Chinese students in America and returned students from America back in China. Unlike the other professional associations organized by Chinese students in the States,[38] the Science Society published a monthly journal called *Science* (*Kexue*) in Shanghai from January of 1915, targeted at a domestic readership to promote general scientific knowledge and increase people's interests in the technical industries in China. It was registered with the Ministry of Education in 1917. After its relocation back to China, the Society purposely recruited other returned students from Europe and Japan in order to expand the Society into a national organization, and the number of members increased gradually from 55 in 1914 to 1005 in 1930.

In addition to receiving a monthly subsidy from the Peking National University for its publication expenses, the Science Society sustained its management by selling its membership as shares of stock initially, and later receiving donations of real estate and funds from the Guandong Provincial Government, the Jiangsu Provincial Government, the China Foundation and later the National Government, and others. In the years 1914–1930, the Society successfully achieved several advances: 1) It set up a Science Library, which was open to the public from 1922; and obtained for it 37,012 volumes in Chinese and foreign languages, and over 140 English, American, German,

French and Japanese journals. 2) It established in 1922 a Biological Laboratory whose two subdivisions of Zoology and Botany attracted much public interest, which led to the creation of the Natural History Museum in Nanjing. After receiving yearly subsidies from the China Foundation from 1926, the research fellows received full-time employment and new equipment, and instruments for physiological experimentation were adequately upgraded. 3) It promoted science education by organizing a summer school at Qinghua University to keep middle-school science teachers up to date, and a Committee on Science Education to investigate the state of technical equipment and science textbooks in the middle schools. 4) It sponsored series of public lectures regularly, and occasionally also talks by western scientists who happened to visit China. 5) It standardized scientific terminology in mathematics, physics, chemistry, and biology. 6) It participated in international science conferences. 7) It set up a Bureau for Scientific Information, an organization which served as a science-policy consultant office for the National Government.

Through its branch office in the United States, the Society maintained its international connection and visibility by attending conferences and exchanging its publications of journals, memoirs, laboratory reports and monographs. Domestically, its eight divisional offices in Shanghai, Beijing, Nanjing, Guanzhou (Canton), Shenyang (Mukden), Hangzhou, Qingdao and Suzhou helped distribute scientific publications and keep up social contact and the mutual exchange of knowledge among members. The Society hosted an annual meeting not only for the election of officers and business discussions in the Executive Committee and Editorial Committee, but also for the annual reports of the Biological Laboratory and the Science Library.[39]

Among other contemporary professional associations such as the Chinese Medical Society, the Chinese Geological Society, the Chinese Agricultural Society, the Meteorological Society of China, the Chinese Society of Engineers, the Chinese Forestry Society and the Institute of Tropical Disease, the Science Society stood out for its unique cross-disciplinary and interdisciplinary character. Different from those of other learned societies, the Science Society's multidirectional goals of "the promotion of science, the encouragement of industry, the establishment of industry, the establishment of translations [of scientific terms], and the dissemination of knowledge" were rather unspecified and ambitious. Nonetheless, as Ren Hongjun recalled in 1937, "We wished to use literary writing to propagate and enlist the support of others, in order to complete the so-called intellectual revolution."[40] We can see from this their conscious attempt to open up a forum in which scientists and non-scientists participated in the definitions and representations of science; hence, their political effects and implications in other social domains were intertwined. The forum of science discourse and other related matters seemed to have a life of their own that resisted any lopsided domination of scientists or non-scientists in the discussion

In the course of science discussion the role of scientists was articulated more clearly, and yet their self-imaging was traced back to that of the previous official-merchant scholar.[41] Even though ideally modeled after the Royal Society of London or the American Association for the Advancement of Science,[42] the Science Society rather grew out of the late Qing era form of voluntary associations (*xuehui*) with a revolutionary and reforming overtone, infused with new elements of professionalism observed from the members' overseas experiences, such as educational qualification, discretionary freedom, control of the right to practice, and corporate self-regulation.[43] What distinguished the Society most from its antecedent *xuehui* was its priority on scientific research and emphasis on experimentation and laboratory investigation as the key to scientific growth. Hence, writing scientific theses according to a particular format – after long hours of laboratory observation, testing, measurement and repeating of experiments – constituted the scientists' praxis and differed from humanistic scholars' self-cultivation in their personal studies and exchange in scholarly gatherings. However, the system of scientific values, or "the scientific ethos," were still very much tied to what Ren Hongjun had termed "intellectual revolution"—a particular relationship of knowledge and action inherited from the Confucian commitment of disinterested but socially-concerned scholarship. It functioned in the traditional society as a counter-current against bureaucratic centralism in a time of peace, and as a reservoir of reforming potential in a time of turmoil.[44]

In comparison, in Japan, as Nakayama Shigeru portrays, occupational supporting groups from the astronomers and interpreters of the 18th century and the physicians of the late 18th-early 19th centuries, to the samurai class of the mid-19th century, played a leading role in the introduction of Western science. After the Meiji Ishin in 1868, Japan had a strong "planned character" in that the government first created institutions for training specific personnel, and only then did university graduates in each discipline organize their scientific societies for the specific purpose of catching up with the West as fast as possible.[45] Far from having scientists follow their own research interests, priority was given collectively to certain basic tasks of state-building. These included geographical and geological surveys, the standardization of weights and measures, meteorological observations sanitation, telecommunication, military works, railways and surveys of natural resources, and were carried out by the Ministries of Technology, Interior and Finance, the Commission for the Colonization of Hokkaidô, the Army, the Navy and other governmental agencies supervising the foreign experts hired in Japan.[46] Accordingly, official control of scholarship was deeply embedded in the institutional process, and the pattern of subordinating scientists and technicians to putative bureaucratic generalists was woven deftly. Scientists were unable to hold any major official posts to participate in upper-level policy-making, which was monopolized by autocratic administrators trained exclusively in law. The treatment of scientists was relatively improved after World War I when the Ministry of Education's Science Research Grants Program and the National Research Council were established in 1918 and 1920, after which time scientists had more political space to maneuver among agencies and could reach out for non-governmental support.[47]

Inasmuch as the State mobilized the institution of various modern disciplines, the government came up with a composite sketch in 1870 for sending students for overseas studies. According to Nakayama, the Japanese educational system from 1872 "borrowed the centralized, pyramidal structure of the French system in externals while being largely American in curriculum and texts." Nonetheless from 1881, in the process of building up the Imperial University in Tokyo, a tilted admiration of German scholarship and higher education became intensified.[48] In the 1870's, only 27 percent of Japanese *ryûgakusei* (overseas students) went to Germany, but in the 1880's it reached 59 percent, in the 1890's 69 percent, and in the 1910's almost 74 percent. Altogether, in the period 1869–1914, two-thirds of the *ryûgakusei* were attracted to study there. As the students officially sent overseas from 1867 to 1920 totaled 3561, we can get an approximate number of 2372 Japanese students who were dispatched to Germany. Discipline-wise, 43 percent of agriculturists took their degrees in Germany, as did 67 percent of chemists, 69 percent of physicists and 90 percent of medical professionals. As Bartholomew's studies reveal, among 1360 holders of Japanese doctorates awarded from 1888 to 1920, 656 (48%) were in medicine, 366 (27%) in engineering, 138 (10%) in agriculture or forestry, 56 (4.1%) in biology, 54 (4.0%) in physics and 90 (6.6%) in other areas.[49] We can not determine at this moment the exact number of Japanese *ryûgakusei* who received their medical degrees in Germany. But we can be sure from the above figures that a predominance of medical doctors were trained by mentors who received their professional training in Germany, and that these two classes formed a vast majority of the Japanese scientific population. Hence, we should not be much surprised by the fact that in Japan the eugenics-campaign activists were mostly physicians and that there were 58 physicians and only 4 biologists who participated in the formation of the Japanese Eugenics Society (*Nihon Yûsei Gakkai*) in 1925.

With respect to the network of the Japanese scientific community, the Tokyo Gakushikaiin (Tokyo Academy), founded under the sponsorship of the Ministry of Education in 1879, institutionally grew out of the Meirokusha (a private society formed to encourage Western learning in 1873) and was modeled after the Royal Society. It later changed its name to Teikoku Gakushiin (the Imperial Academy) and then to Nihon Gakushiin (the Japan Academy) in the post-war era. In a somewhat different way, Japanese specialized academic societies "were typically built around a nucleus of scholars who had studied in the West, and were composed of scientists trained in Japan by foreign teachers" and who were in the process of adopting modern scientific practices.[50] Since the academic structural institution antedated the formation of scholarly circles, these societies were rather like guilds or alumni associations initially. From the guidelines announced by these academic societies, such as the Tokyo Mathematics Association (established in 1878) and the Tokyo Geography Association (1879), they were oriented more toward serving the public than toward scientific research. For *Tôyô gakugei zasshi* (one of the most important scientific journals, which was founded in 1881 by Tokyo Gakushikaiin and modeled after the British journal *Nature*), its principal purpose was to instill scientific ideas and provide better scientific communication.[51] Examining the various medical associations formed after 1874, when the system of physicians was installed, their organizational goals were to exchange medical knowledge and ascertain friendly affiliation and, when necessary, to tackle serious epidemiological problems and public health. Later, these business affiliations were consolidated into a type of physicians union, focused on the enactment of the Medical Practitioner Law to specify physicians' qualifications, status and right to practice. Eventually, the movement reconciled different conflicting interests among regional doctors unions and led to the enactment of the Medical Practitioner Law in 1906.[52]

Comparing the role which scientists played in China and Japan in the institutionalization of biology and medicine, we find a major difference. Whereas Japanese scientists worked like functionaries in the State bureaucracy, Chinese scientists enjoyed the prestigious image of official-merchant scholar and hence certain autonomy. This was despite the fact that both inherited a common cultural legacy from the Confucian commitment of the inseparable relationship of knowledge and action, which values an individual's subjectivity and social responsibility. Ironically, the autonomy of Chinese scientists was later criticized by scientists themselves, in the anti-Japanese mobilization campaigns, as a factor contributing to the difficulty of forming ideologically a national identity and practical team research.

LOCAL SCIENCE AND COMPETITION STRATEGY

Except for the Rinkai jikkenjo (Marine Biological Station of the Imperial University, at Misaki), established in 1887, and the Densenbyô kenkujo (Institute of Infectious Diseases) in 1892, most Japanese research institutes were predominantly the products of war mobilization, founded around and after World War I.[53] The reason that these two research institutes could sustain themselves and advance their international visibility was that their disciplinary grounds were tied closely to local habitats of unique organisms, or endemic contagions. Their research developed into what has been called "local science," a particular approach focusing on problems with a special relationship to or meaning for a specific country or region, as opposed to the more general and universalistic problems that typically occur in fields like physics and chemistry.[54]

This typical strategy of local science escapes the criticism of a lack of imagination, and gains fast international recognition as local scholarship. Such an approach can easily outcompete its alien rivals and lift the previously imposed tag of transplanted science. The strategy was so effective that up to 1885, more than half of the research articles published by faculty members of the College of Science, Tokyo University, were products of local science. Even in universally oriented disciplines such as physics and chemistry, their early scientific concerns were for geophysical observations and the chemical analysis of local products. Disciplines such as zoology, botany and geology remained local sciences until World War II.[55]

In medicine, in addition to the Densenbyô kenkyujo which has been studied very well by James R. Bartholomew, the tendency toward local science more or less predisposed to research activities and social engagement. For anatomy, after an early transitional period of hiring German scientists such as Friedrick Wilhelm Doenits (1833–1912), Hans Paul Bernhard Gierke (1847–1886) and Joseph Disse (1852–1912), Japanese scholars like Koganei Ryôsei and Taguchi Kazuyoshi, who were trained in anatomy and histology by foreign mentors at home and obtained doctorates at Berlin University, started to assume leadership in the Anatomy Institute in the Imperial University from 1885. Their collection of 431 animal skeletons and 1200 human skeletons of different racial-ethnic groups from the stone age turned the Institute into the finest and largest anatomical museum in Japan by 1900.[56] Moreover, the field they established encouraged Japanese doctors to write their own Japanese versions of anatomy, in addition to revising the Dutch medical classic *Kaitai shinsho* (A new text on human anatomy, *Anatomische Tabellen* of Johan A. Kulmus).[57]

Anatomy and pathology were associated with bacteriology and used as effective tools to investigate the etiology of some endemic diseases like dysentery, cholera and plague, as best shown by Kitasato Shibasaburô's discovery of *Yersinia pestis* in 1894, and Shiga Kiyoshi's identification of the first specific etiological agent for bacterial dysentery at Kitasato's Institute of Infectious Diseases.[58] Other pathologic and anatomic studies of the parasitism often found in autopsies were further developed. In 1898, Katsurada Fujirô (1867–1946) and Fujinami Akira (1870–1934) identified a distinct Japanese species of *Distoma*, a liver parasite, that caused syndromes such as swollen liver, bloody stools, nutrimental impediment and abdominal dropsy in humans and other animals, and they subsequently devised a system of treatment and prevention. Later, the parasite's intermediate host was located as the snails often seen in rivers.[59] All these medical discoveries contributed to the application of hygienics, which is by nature a more systematic and aggressive approach to social control and intervention in public health and disease that employs sanitary practices, vaccination, enactment of legislation, public education, social surveillance and other preventive measures. This is so much so that hygienics can be seen as a composite of local sciences.

As Japan transformed itself into an imperialist power and designed its territorial expansion, the local science of biology and medicine became expedient tools for colonial implementation. In 1902, the aforementioned bacteriologist Kitasato Shibasaburô, together with Katayama Kunika (1855–1931), Nagaoka Moriyoshi (1843–1906), Kishida Ginkô (1833–1905), Tanaka Giichi (1864–1929) and Konoe Atsumaro (1863–1904), the president of the Upper House of the Diet, established the Dôjinkai (Association of Benevolence), both to promote close cooperation between Japan and the Asian nations, and to secure regional peace by means of Japanese medical aid and exportation of medical technology. In the 1911 Chinese Revolution, the Association dispatched a medical team of 98 doctors to China, and subsequently took over the management of various hospitals in Manchuria and Korea, before moving on to the Chinese hinterland. In 1924, the Dôjinkai administered the hospital in Qingdao and its affiliated medical school to train Chinese doctors.

In 1925 the Dôjinkai was designated to sponsor, from a partial allocation of the Japanese remission of the Boxer Indemnity, the founding of the Natural Science Research Institute in Shanghai to help produce Chinese doctors and develop Chinese science. The Institute had its opening ceremony in 1931, and included physics, chemistry, biology, geology, pathology and hygiene, bacteriology, and pharmacology in its seven departments, staffed by both Japanese and Chinese research fellows. In 1930, the Dôjinkai established an investigative organ to collect all aspects of information on the public-health system, medical movements, current events and development of endemic diseases in China. In the Association's journal *Dôjin*, there was a special section for circulating medical reports of Chinese endemic diseases and their possible treatments, or plague-prevention activities of the medical teams, various hospitals, and branch offices administered by the Dôjinkai.[60] The Dôjinkai, under the 1939–43 directorship of Konoe Fumimaro (1891–1945, the prime minister in 1937–42 who initiated Japan's expansionist program), comprised 21 important members out of a total of 87 participants in the Japanese Racial Hygiene Association. A well-known member was Koizumi Chikahiko (1884–1945), the minister of the Kôseishô (Ministry of Health and Welfare) in 1941–44, who supported Ishii Shirô's biological weapons research program in Manchuria.[61]

In China, after ten years of effort by the Science Society and financial support from the China Foundation and the Rockefeller Foundation, the Society's Biological Laboratories in Beijing and Nanjing provided academic leadership for the ten biology departments of the five national and five missionary universities, in which they both shared the faculty with the Laboratories' research fellows. Generally speaking, the Biological Laboratories were taxonomically oriented, even though they were simultaneously launching certain projects in genetics, morphology and phylogeny. For instance, Laboratories director Bing Zhi (1886–1965, Ph.D, Cornell, 1918, in zoology), published articles such as "On Some Parts of the Visceral Anatomy of the Porpoise *Neomeris phocoenoides*," "On the Cerebrum of the *Felis tigris*," "Motor Localization of the Cerebral Cortex of the Guinea Pig (*Cavia cobaya*)," and "On Some Points Concerning the Functions of the Mammalian Cerebral Cortex." Chen Zhen (1894–1957, M.A. Columbia, 1921, in genetics) conducted research on the variation and speciation of the Chinese goldfish. He nonetheless emphasized taxonomic studies as the foundation of biological science, arguing that without such studies there was no way to identify the systems of species and no knowledge with which to begin, and that therefore the priority of the Laboratories should be to focus on surveys and collections.[62]

Why prioritize taxonomy? The Science Society's report on its twentieth anniversary pointed out,

> The agenda of our Biological Laboratories in adhering to nation-wide surveys of local flora and fauna in their initial phases was to contribute to the universal system of nomenclature in order to identify myriad species and their distribution in the world. Moreover, the local flora and fauna of a country is precisely the site of the nation's natural resources, and therefore the foundation upon which industry, agriculture, forestry and commerce are able to develop. This is our divine mission as biologists.[63]

In addition to national identity and economic gains, there were other factors predisposing this prioritization. First of all, limited financial resources played a key role, in that the Science Society's Biological Laboratories were barely able to afford more than taxonomic surveys.

Second, although Darwinian evolutionary theory had been introduced into China in the late Qing era, this was the first time that Chinese scientists had converted Darwinism as a theoretical idea into a full-fledged scientific practice, rather than maintain it as mere ideology in the form of Social Darwinism. Taxonomy was looked down upon by some physical scientists and experimental biologists, incorrectly, as simply descriptive activity and unworthy of serious attention. But the study of diversity in taxonomy was the basis for research in major areas of biology. Without it, it would have been impossible for Darwin's *Origin of Species* to unfold the evolutionary history between species and variation among different localities, and the sciences of ethology and ecology would not have been able to develop as they had.[64]

In addition to the above two factors, the following factors were embedded into a nationalist agenda in the production of taxonomic knowledge. China's extensive territory extended from the Cold Temperate to the Tropic zones; it abounded with a wide range of vegetation and endemic species of animals and plants. In addition to these treasures of flora and fauna, there were still untrodden areas needing to be explored in early 20th century China. All these amounted to an irresistible attraction for many European, American and Japanese survey teams dispatched from their home institutes and museums, along with some foreigners who served as faculty in Chinese universities. It is estimated that about 230 foreigners including missionaries, diplomats, military officers and scientists from 16 countries were conducting botanical surveys in China and taking away more than 390,000 floral specimens between 1800 and 1949.[65] Feeling menaced by such encroachments and competition, Chinese biologists were highly motivated to build up their own scientific infrastructure and networks in order to forestall any foreign infiltration. Moreover, the particular interest in producing an ecological account overlapped with the ethnological production of the relationship between people and locality in order to substantiate their territorial claims in the world politics of "geobody."[66]

As with geological surveys, these natural resources needed to be thoroughly identified as to economic values and social functions, and therefore made usable in time of need. The existing literature of traditional medical texts as ready resources added a second incentive for Chinese biologists to pursue taxonomic surveys. On the one hand such surveys moved in the flow of modernization—to scientifically investigate and thereby transform and revive traditional medical culture—and on the other hand they opened up a brave new world of potential scientific medicine.[67] Especially when Japan launched a modernization of Chinese herbal drugs in the 1920s, there was a keen sense of international competition for natural resources and a strong urge for Chinese researchers to produce scientific pharmaceutical drugs through laboratory techniques.[68] For instance, the aforementioned PUMC contributed research on the medicinal use of the Chinese shrub *mahuang* (*Ephedra sinica*): in the early 1920's, when Dr. Carl F. Schmidt and Dr. Chen Kehui in the newly organized Department of Pharmacology isolated ephedrine from this plant.[69]

In medicine, the demand of localization forced PUMC and the Rockefeller Foundation to steer away from the original trajectory of scientific medicine as their medical mission. According to the report of John B. Grant, the professor and head of the Department of Public Health and Preventive Medicine at PUMC from 1921,

> It should be noted that the medical school itself was established chiefly because the founders had decided that their original purpose

> of assisting public health in China was not feasible, due to the absence of a medical profession, and the College was undertaken as a necessary preliminary step looking towards remedying the fundamental deficiency. In fact, so backward were medical conditions that it was decided at first that even a division of hygiene would be temporarily futile. Local conditions fortunately altered this and, by 1924, circumstances permitted the establishment of an independent department [of Public Health and Preventive Medicine] . . . and the Health Station . . . in 1925.[70]

Likewise, the Rockefeller Foundation had undergone in the 1920's a process of reorganization and shift of vision due to a demand for managerial efficiency and turnover of leadership to a new generation. In 1927, the Foundation disbanded the administering organ of PUMC, the China Medical Board, and transferred its function to the Division of Medical Education. In 1928, the Foundation endowed a newly organized American corporation, the China Medical Board Incorporation, with twelve million dollars, and left the new CMB Inc. to manage PUMC's finance independently. With respect to its overall vision, the Foundation shifted from its early commitment to science and research, especially medical research as an end in itself, to new interests in social management and subjects like the social sciences and agriculture. Accordingly, grants no longer went to developmental projects of medical schools, except for projects in research and the advancement of medical knowledge which were conducted by individuals, groups and departments. Moreover, the Foundation became interested from 1931 in the Chinese rural reconstruction movement of James Yen (Yan Yangchu, 1894–1990) in Ding county (Ting Hsien), Hobei. It soon created a China program to combine mass medical education, public health and social sciences into a comprehensive plan embracing both the Foundation's new orientation and the local interest of rural reform on the part of Chinese intellectuals and the government.[71]

In order to expand medical manpower, PUMC accommodated local demands that it support the newly organized Health Ministry of the Nationalist government, after more than a decade of political and military turmoil, to come up with two medical programs, the Commission on Medical Education and the Public Health Training Institute. The Commission urged PUMC to enroll more students, use Chinese as a medium of instruction and focus more on the specific endemic problems of China. The Public Health Training Institute helped reproduce public-health officers and sanitary engineers. In 1936, PUMC also supported the North China Council for Rural Reconstruction in its effort to train university students and faculty from Tsinghua, Nankai, Yenching, and Nanking Universities as medical helpers for rural service.[72]

The adaptation of the Rockefeller Foundation to the native soil as its interests shifted from medical research to social science and agriculture, and the sinification as a result of PUMC's transplantation of the Johns Hopkins model, suggest the gravity of native soil in the implementation of any foreign species of knowledge. Although local inertia has much effect on a science transplanted there, one should not lose sight of the fact that agriculture—as the principal source of livelihood—and public health—as the plan to quarantine and counterattack epidemic and endemic breakouts in the regions — were the long-standing interests of the Rockefeller boards to safeguard their home base in the world of global flows of trade and diseases. Starting in the American South and proceeding to South America, Africa, South Asia and China, the Rockefeller Foundation exported its hybrid model of local/colonial science—special programs of agricultural research and medical education.[73]

EUGENICS AND DISCIPLINING LEARNING

In this section, I explore further the processes of implementing disciplines that not only created the chronological order of these aspects of knowledge but also constituted what and how science should be. Their particular scientific structure, and the local difficulties of limited resources for experiments in order to conform to presumable scientificity, determined an intriguingly cautious positioning of both Japanese and Chinese geneticists toward eugenics and swayed them towards Galtonian biometrics.

Unlike pathology, parasitology, anatomy, botany and zoology, genetics did not come into Japan until the 1890s. Between the 1890s and 1925, genetics research was largely agriculturally oriented, for instance, on rice, silkworms, and wheat as well as Drosophila. Institutionally, this research was mostly produced from the Tokyo Imperial Agriculture University (1890) and Tôhoku Imperial Agriculture University (1911), where the improvement of staple crops was the priority. In 1900, when Mendel's laws were rediscovered after three decades of having been ignored, Tôyama Kametarô, a graduate and then a professor of Tokyo Imperial Agriculture University, interbred the Japanese white-cocoon form of silkworm with French yellow ones. The first hybrid generation was entirely yellow, but its offspring in the next generation were 75.3% yellow and 24.6% white, and backcrossing the first generation hybrids to the white parent produced a proportion of 47.4% yellow and 52.6% white.[74] This experiment was Japan's first extended application of Mendel's laws from peas to animals, and it reconfirmed the simplicity and reproducibility of the first two of Mendel's three laws, that is, 1) the law of segregation, 2) the law of dominance, 3) the law of independent assortment (among loci). This experiment testified that if the factors derived from the parent differed from each other, they would never fuse but invariably segregate during the formation of the germ cells of these hybrids, even though one character was not even expressed in the first hybrid generation. It also kept the segregation ratio 3:1, consistent with a single-particle postulate. Mendel's laws soon were accepted among Japanese geneticists as the genetic basis for the inheritance of discrete traits. It was then suggested by Tôyama, in 1908, that they be used as the basic principles for marriage criteria in order to remove degraded human characters. In 1911, he tried to integrate Galton's biometrics approach with Mendel's genetic laws, under the same framework of hereditary supervenience over environmental factors, into an effective tool with the task of bettering human traits.[75]

Likewise, another horticultural geneticist, Abe Fumio (1879–1945) passionately proposed the use of Mendelism for human improvement without providing a detailed plan of methodology in 1911.[76] However, after 1910 research in genetics was dominated by the Morgan school and was occupied overwhelmingly with purely genetic problems like the nature of the gene and the arrangement of the genes on the chromosomes. And the other Mendelians made little progress in solving the acute problem of the origin of genetic variation.[77] Consequently, Japanese geneticists felt hesitant to jump the gun to conclude Mendelism as an absolute rule to apply in human genetics. Tôyama no longer acted as a spokesperson for Mendelism as the fundamental basis for eugenics, and he died in 1918. As for Abe, in his subsequent efforts in promoting eugenics ideas, he emphasized more of Galton and Karl Pearson's biometrics approach.[78]

During the Taishô era, Nagai Hisomu, the most important key figure in the eugenics movement and the successor of Ôsawa Kenji (1852–1927) as the chair professor of physiology in the Tokyo Imperial University, proposed in 1915 the following technical difficulties in applying the inheritance laws of plants and animals to humans: 1) it is hard to experiment on humans to interbreed appropriately chosen groups; 2) it takes a couple of generations of offspring to determine the genetic results, making it impossible for geneticists to complete testing during their lifetimes; 3) human procreation is far less prevalent than that through the animal and plant kingdoms, and large sample sizes are best for theoretical generalization; 4) it is hard to determine human hereditary units since they are all complexly interrelated in their effect on characters (phenotypic expressions); and 5) in the observation of human genetic diseases, one finds that the symptoms are often manifest in older individuals, so if people die young there is no way to find out what their eventual status would have been, making errors in statistics inevitable. Subsequently Nagai used Galtonian biometrics to accumulate the physical facts of genetic diseases and of physiological heredity predicated upon Mendelian principles, in order to substantiate the claim of a need for eugenic control.[79]

This approach of searching physical evidence with respect to a close connection between genetics and eugenics was further used by Tanaka Yoshimaro (1884–1972), another expert on silkworm genetics and the founder of the Japanese Genetics Society. In 1920, he made a fervent plea for the establishment of a eugenics institute whose main tasks would be: 1) nationwide genetic surveys of heredity and specific characters, 2) comparative studies of eugenics in Japan and other countries, 3) investigations of divorce results, 4) experimental genetic studies, 5) preventing any breeding of genetic defectives, 6) devising methods of encouraging the propagation of superior elements, and 7) spreading eugenic ideas.[80] The desire to elevate genetics into an institute-based discipline impelled other geneticists such as Miyake Kiichi, Komai Taku, Kawakami Riichi and Furuhata Tanemoto to participate enthusiastically in the eugenics movement and to start working on human genetics in order to promote wide support for such an institute.[81]

Komai's series of genetic studies, starting from his initial report in 1911 on two cases of albinism in a family tree in a village of Kagoshima prefecture, were much influenced by his mentor Oka Asajirô's biometrics. And he often consulted for reference the works of Charles Davenport, among whose intellectual influences were Ronald A. Fisher, Karl Pearson and Francis Galton. Kawakami, although trained in medicine, was the pioneer and activist for the founding of the Jinrui idengaku kenkyûkai (Society for human genetics studies) in 1940. His research was focused on genetic aspects of colorblindness and kin marriage. He expounded his basically biometrics-oriented methodology, and it was adopted by most of his colleagues in the Jinrui idengaku kenkyûkai for the purpose of collecting various data on genetic diseases as a theoretical basis for eugenics. Furuhata was a serologist and in 1925 started an investigation of 459 people in 101 families to find the connection between blood types and hereditary succession. The biometrics approach became very popular among members of the Japanese Racial Hygiene Association (1930) such as Furuya Yoshio, Itô Makoto and Imamura Shôichi, who earnestly advocated it as the scientific basis for racial biology.[82]

In addition to the apparent methodological obstacles confronting the human application of Mendelism, the specific choice of Galtonian biometrics was related to the very structure of medicine and especially biology. This underlying structure constituted what and how science should be. As prominent philosopher Tanabe Hajime (1885–1962) pointed out in his book *Saikin no shizen kagaku* (The Recent Natural Sciences) in 1915, the processes of scientific methods have the goal of abstracting universal elements from myriad experiences into simple concepts and therein to manifest generalization. In order to infer universal elements, one needs first to analyze these experiences; therefore the two levels are translatable and partly interchangeable. Scientific research takes the process a step further, not only analyzing the common sense but also defining the form and content of scientific concepts so as to encompass all possible kinds of experience and allocate them into appropriate categories of classification. To come up with a system of classification, one needs accurate observation and penetrating analysis to discern the character and nature of objects. Finally, one uses a language of description to present objects of the same category. Naming and ordering are actual practices in descriptive aspects of sciences like zoology, botany, and mineralogy. However, to derive universal laws and generalizations from these categorized experiences one has to move forward to deduction—establishing a hypothesis and deducing its predictions. If the predictions turn out to be incorrect, the hypothesis has to be modified to conform to actual experience. An adequately verified hypothesis has the status of a law. Still, the law must be further tested to see whether it holds in other contexts, such as conditions in China and Japan. With a view to complete the processes of knowledge in natural science, one has to constitute the law of evolution in order to explain the origin of species, and the descriptive aspects of zoology and botany should move on to so-called "explicative science" such as physiology, cytology and morphology. Both descriptive and explicative sciences are indispensable to each other in terms of value and solidity of knowledge.[83]

The above abstract of Tanabe's statement is not to reiterate what science thereafter should be, but rather to demonstrate a perception of scientificity parallel to the hierarchical structure of the explicative over the descriptive and the chronology of the descriptive prior to the explicative, implemented around the same time. It is important to note that what Japanese science transplanted from the West was not merely universal laws, but also mechanisms of knowledge production—the capacity to undergo the very scientific processes which would in turn test "universal" laws. In this sense, local science can produce results which force proposed general laws to be modified. A helical progress of understanding thus results with local experience and theory each modifying the other sequentially. Despite their propaganda of eugenic ideas, Japanese geneticists sensed a pressing need for serious studies of eugenics from the ground up, there being no suitable data to speak of. Hence, it should not be surprising that they emphasized biometrics so as to gather an avalanche of numerical data from which scientists could attempt to locate problems and devise plans of solution.

The same pattern of scientific processes can be observed in the Chinese context with an even more aggravated intensity, since there was a heated debate in the 1930s as to what science should be, conducted by the experimental and the survey-investigation cliques (*shiyan pai* versus *diaocha pai*) in Chinese biology. In spite of the aforementioned factors predisposing the priority of taxonomy and the research projects of survey and data collection, some biologists challenged this particular preference. The debate was instigated in 1932 by Wang Jingxi (1893–1968, Ph.D, Johns Hopkins, 1923, in physiology and psychology), who criticized Chinese biology for falling behind and going against the current of the world's experimental development of genetics, embryology and physiology. He said,

> Taxonomy is going downhill, and even morphology has begun to adopt experimental methods. Over the years, Chinese biology has spent most of its funding and human resources on taxonomy and morphological surveys, and this has resulted in a lopsided and unhealthy growth. Now, it is about time to promote experimental biology. Otherwise we would end up losing a vision of the world and focusing merely on the limited interests of a small number of scientists.[84]

The botanist Wang Zhenru (1908–?) from the Science Society countered that the reality of application was the key concern for Chinese biology, and that Chinese scientists should face the actual difficulty of building infrastructure—biological surveys and basic nomenclature – instead of merely looking at one's personal interests. He used a statistical survey of the 1927 and 1930 volumes of Biological Abstracts to support his point that taxonomy and morphological identification were still trendy practices in Western botany. By virtue of the reality that financial support for biological research over the vast territories of China was seriously inadequate, the responsibility of scientists was not to discourage the already-deprived discipline but to help arouse public recognition and support for more taxonomic research activities.[85]

Another botanist, Hu Xiansu (1894–1968, Sc.D, Harvard, 1925, botany), who was one of the founders of the Science Society and who took charge of the botanical section of the Biological Research Laboratory, espoused Wang Zhenru's position and argued powerfully,

> Biology is different from other universalistic disciplines like physics and chemistry, which can be transplanted readily without too much advance preparation. Although Chinese scientists receive systematic training in taxonomy in the West, once they return to China they have to start all over from scratch because different species occur in different geographical zones. . . . Over the years of effort by taxonomists working under the difficulty of scarce resources, we have earned international respect for our independent scholarship. The Chinese flora has become our own specialty among world botanists. How short-sighted the vision and insufficient the common sense of scholars who complain about an over-development of taxonomy in Chinese biology.

Moreover, he argued that the taxonomy in Wang Jingxi's mind was the already-outmoded typological system of classification, and that Wang was ignorant of the progress made in botany since the 19th century in which the experimental methods of animal serology, protein identification and chromosome numbers became accepted practices to determine the relationships of genera. He predicted that in the future the combination of experimental methods and morphology would be unquestionably the guiding approach in taxonomy, and that Chinese scientists needed to do their very best to catch up.[86]

Both camps in their subsequent articles reiterated the above positions. Ostensibly they were arguing about two different methodological approaches—descriptive and experimental—as being the most appropriate for Chinese biology at the time; however, I do not see any fundamental conflicts in their perceptions of science. Even though they disagreed on the priority of financial allocation and different evaluations of timing in research development, each re-confirmed the other as to scientific processes—the temporal priority of surveys, identification, naming and classification over experiments.

This perception of scientific processes resonated with a popular exposition of scientific methodology around this period, often under titles like *Kexue fangfa* (Scientific Methods), *Kexue zonglun* (Introduction to Science) or *Kexue Gailun* (General Discussion of Science), that more or less resembled Tanabe's writing.[87] This disciplining force impelled both Chinese geneticists and eugenicists to measure up to the well-accepted scientific standard. Geneticists, whenever they touched upon human heredity and inherited diseases, were implicitly presenting evidence to support human improvement. They presented the eugenic possibility on the basis of successful results in crop production, and they often used Galtonian genealogical statistics of inferior traits or desirable characters (mostly social achievement and status) among certain families to demonstrate their hereditary connection. For instance, the experimental geneticist Chen Zhen wrote "Yichuan yu wenhua (Heredity and Culture)" in the journal *Kexue* (Science) in 1922,

> Because of the fact of heredity all men are created unequal. Sometimes we see genius in a poor family or mentally-retarded progeny with decent parents, and we quickly conclude that heredity is not a congruous explanation for the observed phenomenon. However, if we track down their genealogical records for about ten generations and provide a series of statistical accounts, we can solve the puzzling incongruity.

Chen Zhen provided a genealogical comparison of three lineages, the Jukes, the Jonathan Edwards, and the Kallikaks in America, to sustain his claim that heredity but not environment was the dominant factor in determining the rise and decline of a family and, by extension, a nation and a culture. And then Chen had to muffle his assertive tone to ask his readers not to mistake him to be saying that environment was irrelevant or unimportant. Chen appealed to Galton's study concluding that the inherited traits of his British contemporaries were no match for those of their talented ancestors because the significance of heredity in domestic breeding had not been recognized in the society in previous centuries. Chen said that the First World War was a bloody living example of not taking the fundamental approach of genetic improvement, and reiterated that it was about time to turn to genetics for a change. Because the physiological, embryological, and phylogenetic aspects of humans did not differ from those of other plants and animals, and evidence of genetic manipulation had already been shown in crop and animal betterment, Chen saw a possibility for a eugenic control of social ills. As a geneticist, Chen concluded his article with a possible connection between genetics and eugenics, and nothing further.[88]

Another biologist, Chen Jianshang (1898–1988, a Science Society member and student of the geologists Wong Wenhao and Ding Wenjiang, with a specialty in ichthyological taxonomy), wrote a chapter on "Yichuan yu rensheng (Heredity and human life)." In it, he accentuated the importance of applying genetic laws to human life, given the agricultural achievement of such application. Like Chen Zhen, he used the statistical example of the American Jukes lineage, the majority of whose offspring were regarded as socially unfit, and he also made a contrasting model with the eminent English Darwin family, in which 16 scientists were born, in order to press the claim of genetic improvement as a key to social reforms. He said that Francis Galton was the first person to apply scientific methods to human improvement and establish a research institute for this specific purpose, and in China such kinds of trendy activities were already being reported in the media. For eugenic manipulation there were positive and negative measures, the former to encourage breeding among the fit, and the latter the opposite. Despite his apparent promotion of eugenic ideas, he concluded that it was extremely difficult to put them into practice, since the fact has shown that the eugenics movements in various countries did not attain the expected results. "Maybe!" he said. If people believed such eugenic ideas and turned them into a national program, the outcome might be different.[89]

Without any basis in scientific investigation of degeneration problems in the native soil, the two biologists could present a mere possibility for implementation of eugenic methods for human improvement. Unlike the active participation in the West, this cautious positioning of Chinese geneticists toward eugenics was a professional reaction to a publicly interesting topic and a not yet thoroughly examined subject. To make it more scientifically sound, the eugenicist Pan Guangdan changed his writing format in the 1930's. Scrutinizing his articles of the 1920's collected into a volume entitled *Youshen gailun* (Essays on eugenics), I notice that they were rather informative reports on current events or historical accounts of the eugenics movements outside China.[90] From 1927, a year after he returned to China from Columbia University, we start to see more engagement of local issues in Pan's writing.[91] In 1930, his column "Xuedeng" in *Shishi xinbao* (Current News) was published in a collection. In it, his short essay "Kexueyanjiu yu kexue tichang (Scientific research and science propagation)" expressed his deep disappointment about some of the Science Society leaders lingering on the stage of science propagation without putting it into action and launching needed scientific research. He asserted that the only way to popularize science and make it respectable was to do research. Among his articles and ongoing disputes with other intellectuals on eugenics-related issues like marriage laws, intelligence tests, national characters and the family system, he did conduct a questionnaire survey on the Chinese family system. In it, he addressed issues such as ancestor worship, the family system, genealogical records, purposes of marriage, criteria for marriage partners, marriage age, kinship, concubinage, divorce, birth control, number of children, and the rearing and education of children. Pan received 317 replies to his questionnaire, 273 from men and 44 from women, and he subsequently transformed them into statistical tables and attached his own long analysis.[92] In his 1930 article "Renwen xuanze yu zhonghua minzu" (Cultural selection and the Chinese race), in which he discussed lineage and the traditional family system, and the electoral system as a function of cultural selection. To this article he attached a questionnaire entitled "A study of the quality of Han immigrants in Manchuria," although we do not have his follow-up report on the outcome of this research.[93] All his efforts here were to produce statistical data or transform historical materials into accountable statistical data with which he could meticulously present a problematic phenomenon and push forth a problem-solving eugenics proposal. The specific details of his eugenics proposals will be discussed in the next chapter.

Pan's genealogical studies of traditional Chinese theatrical performers (*Zhongguo lingren xueyuan zhi yanjiu*, 1934) and of respected families in Jiaxing during the Ming and Qing periods (*Ming-Qing liangdai Jiaxing de wangzu*, 1937), and the table of family trees of historical figures (*Cunren shuwu lishi renwu shixi biaogao*, 1937–38) exemplified such an approach. Without being too detailed about each of his works, in the following I will just sketch Pan's approach in *Zhongguo lingren xueyuan zhi yanjiu*. Initially, this work was reminiscent of Galton's *Hereditary Genius*, as Pan cited Galton's brief discussion of the performers' genealogical background. Nevertheless, Pan's methodological frame of reference was rather a composite one—an approach he conceived partly from the inspiration of his mentor Charles Davenport who, rather than sharing Galton's and Pearson's concern with the "phenotype," the expressed and therefore observable characters, was interested in the "genotype"—the individual's genetic makeup that "had to be inferred from scrutiny of as many related phenotypes as possible, in and beyond the immediate family."[94] In addition to Davenport's emphasis on the collection of extended family pedigrees, Pan was inspired as well by both Havelock Ellis's *A Study of British Genius* and Montrose J. Moses's *Famous Actor-Families in America*, in which the genres of biography and genealogical succession were used as forms of narrative that could elaborate the details of distribution of nationality, social status/mobility, skin color, race crossing, and lineage establishment. With this composite approach, Pan collected not only official biographies and private accounts of actors/actresses throughout different historical periods, but also their family histories and genealogical records. He explored the historical and geographical distribution and migration of actors/actresses, the kinship within the network of career associates, the succession of particular acting roles within lineages, and their class distribution and social status, in order to fit the genealogical into the Mendelian frame.

The particular choice Pan made, to do research on historical figures and their genealogical succession to illuminate the scientific base of eugenics and meet the disciplinary demand of scientificity, was determined in part by logistic constraints: 1) Official biographies and family genealogies are relatively reliable and easily available resources in China, so that the renowned geologist Ding Wenjiang had advocated in 1919 the statistical study of these materials to convince Chinese readers of the scientifically-based knowledge of eugenics.[95] 2) Unlike his American mentor Davenport, who received a generous endowment for his research from Andrew Carnegie, Pan hardly received any grants and was forced to launch small-scale research projects, and hence for him the most efficient way was to transform pre-existing materials into usable data.

CONCLUDING REMARKS

I have discussed, in a global context, the American and Nazi connections of the eugenics movement, as well as the formation of the scientific community and disciplinary establishment of biology and medicine in both Japan and China in which eugenicists struggled for social recognition and disciplinary turf. With limited resources, their competition strategy of converting local materials with modern scientific language into usable data in order to fast gain recognition, resonated with other local scientists in biology and medicine. In other words, Pan Guangdan's decoding and converting of conventional family genealogies paralleled the translation of local gazetteers by the Japanese and Chinese geologists, geographers and ethnologists. This strategy of turning language barriers into a national fortress allowed local scientists to compete with their alien rivalries and forestall foreign infiltration.

The particular scientific structure and perception of scientificity determined a cautious positioning of both Japanese and Chinese geneticists toward eugenics in the 1920s and early 1930s, and forced the eugenicists lingering over the Galtonian biometric approach to data collection and evaluation before they could come up with a eugenic plan and enactment of social control. Consequently, Pan made a discipline choice of shifting from biology to sociology in his career. Likewise, the Japanese eugenicists, mostly physicians, who were the predominant majority of the scientific community, had to measure up to the same standard of scientific validity. Thus, when the

Japanese Eugenics Association (*Nihon Yûsei Gakkai*) was founded in 1925, the Association set up five branch departments for scholarship, documenting, publishing, counseling and finance, in order to facilitate survey and investigation projects of physical, mental and hereditary unfitness, documentation of the Japanese national character, and propagation of eugenic ideas. The department of scholarship was designed to seek collaboration of expértise from various fields such as anthropology, physiology, psychology, genetics, comparative anatomy, biology, experimental philosophy, psychiatry, criminology, statistics and sociology.[96] In 1930, when the Japanese Racial Hygiene Association (*Minzoku Eisei Gakkai*) was established, the Association's prospectus declared significantly that the most important task of the racial-hygiene study was the fundamental, especially biometric, investigation of the Japanese physical constitution and heredity. As an institution for this particular purpose of enlightenment, the Association's journal was aimed at circulating results and reports of such surveys and investigation.[97]

NOTES

[1] Yu Fenggao, *Lu Xun zawen yu kexueshi* (Lu Xun's essays and history of science) (Hangzhou: Zhejian wenyi, 1986), pp. 151–3.
[2] Arjun Appadurai, *Modernity at Large* (Minneapolis: University of Minnesota Press, 1996), p. 2.
[3] See Prasenjit Duara "Imperial Nations/National Empires: Ethnography and the Territorial Imperative in East Asia," unpublished paper.
[4] Suzuki Zenji, "Geneticists and the Eugenics Movement in Japan," pp. 157–164.
[5] Ibid. When the Japanese Racial Hygiene Society (*Minzoku Eisei Gakkai*), based in the Department of Physiology at Tokyo Imperial University, was established in 1930, it was guided mainly by physicians and led by Nagai Hisomu (1876–1957), a medical doctor at Tokyo Imperial University. It was at about this period that the Japanese eugenicists began to select German eugenics as a model.
[6] Suzuki, p. 163.
[7] Ibid., pp. 158–159.
[8] Laurence A. Schneider's "Genetics in Republican China" in *Science and Medicine in Twentieth-Century China: Research and Education*, (Ann Arbor: University of Michigan Press, 1988), ed. by John Bowers et al., pp. 3–29.
[9] Ibid.
[10] The quotation is excerpted from Donna Haraway's *Primate Visions: Gender, Race and Nature in the World of Modern Science* (London: Routledge, 1989), p. 135. Although Haraway does not provide the source of her statement, Lily E. Kay's *The Molecular Vision of Life:: Caltech, the Rockefeller Foundation, and the Rise of the New Biology* (Oxford: Oxford University Press, 1993) is a good reference work on the Rockefeller Foundation's creation of the molecular biology program at the California Institute of Technology, in addition to its support for the University of Chicago, Stanford, Columbia, Harvard and the University of Wisconsin. "From 1932–1959, the Foundation poured about 25 million into the molecular biology program in the United States, more than one-fourth of the Foundation's total

spending for the biological sciences outside of medicine....[T]he pre-World War II level of Rockefeller Foundation support for molecular biology amounted on average to about two percent of the entire federal budget for scientific research and development." (p. 6) Before its molecular biology program of 1932, the RF had financed biology for 15 years. From here, we can come up with an approximate figure of $100 million of Biology expenditure by the RF between 1917 and 1945.

[11] E. Richard Brown tries to make a strong case that the RF activities in China were, by definition, "imperialism," but he fails to take account of the subjectivity of the local scientists and their interaction with colonial science, which can be characterized as "local science" that has a transforming potential to assimilate it and turn it into a different colonial version. See his article "Rockefeller Medicine in China: Professionalism and Imperialism" in *Philanthropy and Cultural Imperialism: The Foundations at Home and Abroad*, ed. by Robert F. Arnove (Boston: G.K. Hall & Co., 1980), pp. 123–146.

[12] See Proctor's *Racial Hygiene—Medicine Under the Nazis*, (Cambridge: Harvard University Press, 1988), pp. 97–101.

[13] Stefan Kuhl, *The Nazi Connection—Eugenics, American Racism and German National Socialism*, (New York: Oxford University Press, 1994), pp. 53–63.

[14] Ibid., pp. 65–76.

[15] Ibid., pp. 97–104. These American prosecutors at Nuremberg not only tracked down the guilt of German doctors but also were involved in recruiting for American military research. See Christian Pross, "Medicine without Compassion" in *The Nazi Doctors and the Nuremberg Code*, ed by George Annas and Michael Grodin, (New York: Oxford University Press, 1992) , 32–52.

[16] The Rockefeller grants to Demark totaled about $90,000. The major funding the Rockefeller Foundation granted the Kaiser Wilhelm Institute was $325,000 for the construction of a new building, not to mention other individual fellowships. See Kuhl's *Nazi Connection*, p. 20. The sum of grant to the Institute could be even larger, as seen from Paul Weindling's "The Rockefeller Foundation and German Biomedical Sciences, 1920–40: from Educational Philanthropy to International Science Policy," in *Science, Politics and the Public Good—Essays in Honour of Margaret Gowing*, ed. by Nicolaas A. Rupke, (London: the Macmilan Press, 1988), especially p. 131. According to Proctor's account, the Rockefeller Foundation provided funds to the Fischer Institute for twin studies until the mid—1930s, and when the foundation stopped this funding, the American philantropist James Loeb gave $1 million to the institute to continue research in this area. By 1940, the Institute research fellows had produced a total of 569 scholarly works on racial hygiene and related topics. See *Racial Hygiene*, pp, 349n.115, 40–41.

[17] Raymond B. Fosdick, *The Story of the Rockefeller Foundation*, (New York: Harper & Brothers, 1952), pp. 88–89.

[18] Laurence A. Schneider, "Genetics in Republic China" in *Science and Medicine in Twentieth-Century China: Research and Education*, (Ann Arbor: University of Michigan Press, 1988), ed. by John Bowers et al., pp. 3–29. And Juliette Chung, "Struggle for National Survival: Chinese Eugenics in a Transnational Context, 1896–1945" (Dissertation, University of Chicago, 1999), chapter III.

[19] Ibid., p. 139.

[20] See Sumiko Otsubo Sitawich, "Eugenics in Imperial Japan: Some Ironies of Modernity, 1883–1945," (Dissertation, Ohio State University, 1998), chapter V. Hiratsuka Raichô and Gotô funded the campaigns from their personal pocket money; the educator Naruse Jinzô raised public funds to build a women's university, within which he planned to set up a department of racial improvement; Nagai Hisomu relied heavily on membership revenue to finance eugenics research and public campaigns; only a few individuals, such as Yamanouchi Shigeo, were able to conduct research on the fellowships from the connections of Columbia University, the University of Chicago and the Carnegie Institution.

[21] The Chinese eugenicist Pan Guangdan was a student of Charles Davenport at Columbia University and had kept a close relationship with Davenport while developing his career in China. Sumiko Otsubo and I discuss the impact of the American 1924 Immigration Restriction Laws by Chinese and Japanese eugenicists in our dissertations.

[22] Ka-che Yip's *Health and National Reconstruction in Nationalist China: The Development of Modern Health Service, 1928–1937* (Ann Arbor: AAS), 1995, p. 13.

[23] Fu Weikang's *Zhongguo yixueshi* (Shanghai: Shanghai Zhongyi, 1990), p. 449. Although there were missionary medical colleges existing prior to this period, they merely produced medical helpers and assistants through apprenticeship.

[24] Frank Ninkovick, "The Rockefeller Foundation, China, and Cultural Change" in *The Journal of American History*, 1984, 70:4, pp. 799–820.

[25] Qiusha Ma, *The Rockefeller Foundation and Modern Medical Education in China, 1915–1951* (Dissertation, Case Western Reserve University, 1995), chapter III: "China: A New Frontier for the Rockefeller Philanthropy."

[26] Mary B. Bullock, *An American Transplant: The Rockefeller Foundation and Peking Union Medical College* (Berkeley: University of California Press, 1980), chapter II: "A Johns Hopkins for China."

[27] Ibid., and Qiusha Ma's, chapter IV: "Medical Education as An Approach to China: Divergence Between the Rockefeller Foundation and Missionaries."

[28] Ibid.

[29] Up to 1942 PUMC had produced 2288 doctors, missionary or Chinese, in their advanced training, with 315 Ph.D. degree holders among them. See Ibid., p. 164.

[30] Schneider's "The Rockefeller Foundation, the China Foundation, and the Development of Modern Science in China" in *Social Science and Medicine* 16 (1982): 1217–21.

[31] Ibid., and for a detailed discussion of the establishment of the China Foundation and its reorganization, see Yang Tuihua's "Zhongjihui de chengli yu gaizu" in *Jindaishi yanjiuso jikan* 18 (Taipei: Academia Sinica, 1989), pp. 259–280.

[32] Ren Hongjun, "Shinianlai Zhongjihui shiye de huigu"(Review of the China Foundation's ten-year achievements) in *Dongfang zazhi*, vol. 32, no. 7, 1935, pp. 19–25.

[33] Ibid.

[34] Ibid.

[35] Liu Zhaomin, *Zhonghua shengwu xueshi* (The history of Chinese biology) (Taipei: Shangwu, 1991), pp. 463–464.

[36] Some trustees on the board were also members of the Science Society, for instance, Cai Yuanpei, Wang Jingwei, Sun Ke, Wong Wenhao, Zhao Yuanren, Hu

Shi and Ren Hongjun. In particular, Ren became the Executive Director of the Foundation. For a detailed discussion of the restructuring of the China Foundation and its political struggle, see Yang Tuihua's "Zhongjihui de chengli yu gaizu."

[37] As noted before, Ren became the Executive Director of the China Foundation. Hu Shi was a leader of literary movements in the 1920's and 1930's; Zhao became an internationally renowned linguist; Hu was a professor of mathematics at the Datong University at Shanghai; Ping became a professor at the University of Nanking and director of the Science Society's Biological Research Laboratory; Zhou was the first director of Academia Sinica's Research Institute of Engineering; Yang became General Secretary of Academia Sinica; Guo was a professor of agriculture at the National Southeast University; and Jin became the Chancellor first of the Peking Government Agricultural College and later of Qinghua University. See Peter Buck, *American Science and Modern China, 1876–1936* (Cambridge: Cambridge Univ. Press, 1980), pp. 94–95.

[38] According to Weili Ye's studies, there were at least 200 learned societies formed in the 1870's and 1880's, and from the 1890's to 1910's there was an active organization called the Shi Shao Society, and yet like the previous ones, it bore the character of the study-group movement. It was not until the 1910's that a sense of professional identity became to emerge. During this period, there were the Chinese Academy of Arts and Science with 100 members in its initial year, the Chinese Engineering Students' Society, the Engineering Committee of the CSA, the Arts and Sciences Group, the Agriculture and Forestry Groups, and the Science Society of China. See the second chapter of her *Crossing the Cultures: The Experience of Chinese Students in the U.S.A. 1900–1925* (Dissertation, Yale University, 1989).

[39] This brief history is based on the Society's own publication, *The Science Society of China: Its History, Organization, and Activities* (Shanghai: the Science Press, 1931).

[40] David C. Reynolds, *The Advancement of Knowledge and the Enrichment of Life: the Science Society of China and the Understanding of Science in the Early Republic, 1914–1930* (Dissertation, University of Wisconsin-Madison, 1986), p. 25.

[41] Peter Buck did a sampling study of the scientists' birthplaces and their relations to economic development, historical academic performance and Christianity conversion rate. According to his analysis, most scientists in his sample were born and raised in the well-off counties of Zhejiang and Jiangsu provinces, where their social and political achievements were markedly above those from other areas, and their conversion rate was relatively low even though these places were exposed to missionary science for easy accessibility of transportation. In addition, the Science Society was rather a mixed evocation of conventional chambers of commerce and voluntary associations. See his *American Science and Modern China*, Chapter IV: Science and Revolution: China in 1911.

[42] Ibid., p. 116.

[43] See Joseph Ben-David's better elaboration of professionalism in his "The Profession of Science and Its Power" in *Essays on the Social Organization and Ethos of Science* (Berkeley: Univ. of California, 1991), pp. 187–209.

[44] *The Advancement of Knowledge and the Enrichment of Life*, Chapter II: "Scholarship, Association, and the Science Society of China, 1914–1919."

[45] Shigeru Nakayama, *Academic and Scientific Traditions in China, Japan, and the West*, trans. by Jerry Dusenbury (Tokyo: Univ. of Tokyo Press, 1984), Chapter VI: "The Transplantation of Modern Science," pp. 193–237. Most early Meiji scientists came from the samurai, and had fathers with strong connections to Confucian scholars, Japanese mathematicians, scholars of Dutch studies, or Chinese-medicine doctors. See James R. Bartholomew's *The Formation of Science in Japan: Building a Research Tradition* (New Haven: Yale Univ. Press, 1989), Chapter IX: "Science and Society: A Retrospective."

[46] Nakayama, pp. 210–211.

[47] Bartholomew, pp. 268–272.

[48] Nakayama, p. 219. For the subjects of physics, astronomy, geology and mineralogy, chemistry, zoology and botany, medicine, pharmacology, educational system, political science, and economics, students would be dispatched to Germany. There were some overlaps in certain subjects. In Britain, Japanese students would study machinery, geology and mining, steel making, architecture, shipbuilding, cattle raising, commerce, poor-relief; in France, zoology and botany, astronomy, mathematics, physics, chemistry, architecture, law, international relations, promotion of public welfare; in Holland, irrigation, architecture, shipbuilding, political science, economics, poor-relief; and in the U.S.A., industrial laws, agriculture, cattle raising, mining, communications, commercial law.

[49] See Bartholomew's Chapter IV: "Formation of the Meiji Scientific Community." In addition to the intellectual appreciation of German scholarship, other factors like its easy accessibility of scholarly information in the academic systems and its willingness to relax the special privileges accorded its nationals under the system of unequal treaties (1857–99) appeared much more attractive. Also, Bartholomew shows that some scientists complained of a tilted balance in that manpower, resources and action were in medicine and other fields got what was left; see p. 274.

[50] Nakayama, p. 213. Nakayama describes this as a phenomenon of the professional "chair system," which proved both opportune and efficient in the deployment of Modern Science. This system explains the close bond between mentors and pupils. Unexpectedly, this also created factionalism that impeded communication, and criticism among the scientific communities, even though it served the function of stimulating competition among the research institutes and universities. See also James R. Bartholomew's Chapters VI: "Scientific Research in its Social Setting," and IX: "Science and Society: A Retrospective."

[51] Horiuchi Gôji, *Kindai kagaku shisô no keifu* (The Genealogy of modern scientific thought), Tokyo: Shibundô, 1964, pp. 181–182.

[52] Shugaya Akira, *Nihon iryô seidoshi* (The History of Japanese medical institutions) (Tokyo: Hara shobô), 1976, 240–260. The renowned bacteriologist Kitasato Shibasaburô, studied by James R. Bartholomew, was one of the movement's leaders. (Despite Bartholomew's use, people from Kitasato's hometown in Kumamoto called him "Kitazato").

[53] Nakayama Shigeru, *Characteristics of Scientific Development in Japan* (New Delhi: The Center for the Study of Science, Technology and Development, 1977), p. 47.

[54] See his *Academic and Scientific Traditions in China, Japan, and the West*, p. 220. Local science, according to Nakayama, is intended to be a further elaboration and development of normal science that a system of modern scientific methods and disciplines introduced onto non-Western soil, which pre-modern investigation was predominately practiced. However, as a result, it underwent an intriguing transformation that transfigured both implanted and pre-existing local knowledge. The transformation of local science might turn out to challenge the authority of paradigmatic science originating in the West. In other words, the original has no control over what its copy will turn out to be.
[55] Nakayama's, *Characteristics of Scientific Development in Japan*, p. 48. Another factor of local science inclination has to be considered in that, like China, there was an abundant literature of taxonomy in Japanese traditional biology.
[56] John Z. Bower, *When the Twain Meet: The Rise of Western Medicine in Japan* (Baltimore: The Johns Hopkins Univ. Press, 1980), pp. 98–99.
[57] *Monogatari Nihon kagakushi*, p. 338. The Japanese translation was revised once in 1826.
[58] For the story on why the plague bacillus was named not after Kitasato but after Alexandre Jean-Emil Yersin (1863–1943), see Bower, pp. 112–114. For Kitasato's struggle with the bureaucracy in the process of discovery, see Bartholomew, Chapter V: "Science and Bureaucracy."
[59] *Monogatari Nihon kagakushi*, pp. 342–343
[60] See *Dôjinkai yonjûnen shi* (The Forty years of the Benevolence Association), ed. by Hosaka Yûiichirô (Tokyo: Dôjinkai, 1943), and Huang Fuqing's *Jindai Riben zaihua wenhua ji shehui shiye zhi yanjiu* (The Study of modern Japanese cultural and social enterprises in China) (Taipei: Academic Sinica, 1982), Chapters II and III.
[61] See *Dôjinkai yonjûnen shi*, pp. 24–26 and also the journal *Minzoku eisei*, vol. 4, no. 1, pp. 87–101, for the Racial Hygiene Association membership namelist. I compared the membership list with that of the Japanese Racial Hygiene Association to come up the number of overlapping members.
[62] See Chen's biography in *Zhongguo xiandai kexuejia zhuanji* (Biographies of modern Chinese scientists) (Bejing: Kexue chubanshe, 1991), vol. 1, pp. 458–468, and his article "Guonei Shengwu kexue jinnianlai zhi jinzhan (The recent development of our biological science)", in *Dongfang zazhi*, vol. 28, no. 13, 1931, pp. 99–110.
[63] In the Society's journal *Kexue* (Science), vol. 26, no. 1, 1944.
[64] See Ernst Mayr, *The Growth of Biological Thought: Diversity, Evolution, and Inheritance* (Cambridge: Harvard University Press, 1982), pp. 142–143.
[65] See Chen Temao, *Zhongguo zhiwu fenlei xueshi* (History of Chinese botanical taxonomy) (Wuchang: Huazhong shifan, 1993), pp. 175–180.
[66] The term "geobody" was coined by Thongchai Winichakul in his *Siam Mapped: A History of the Geo-Body of a Nation* (Honolulu: University of Hawai'I Press, 1994). For an elaboration of "geobody" in the Chinese and Japanese contexts, see Prasenjit Duara "Imperial Nations/National Empires." Also see Fan Fa-ti, "Nature and National Essence: the *Guocui* Circle in Early Twentieth-Century China" in *The Moral Authority of Nature*, ed. by Lorraine Daston and Fenando Vidal, (forthcoming, Univ. of Chicago Press).

[67] See Wu Liangde's "Zhongguo yixue zhi fuxing (The Revival of Chinese medicine)" in *Kexue*, 1935, vol. 20, no. 4, pp. 259–266. There was also a survey report, posted in *Kexue*, of Chinese herbal medicine in Sichuan which advocated more activities of such survey projects; see Xu Zhifang's "Zai shu souji minjian sheng caoyao zhi chubu baogao, in *Kexue*, 1933, vol. 18, no. 1, pp. 28–31. Also see Hsiang-lin Lei, "When Chinese Medicine Encountered the State: 1910–1949," (Dissertation, University of Chicago, 1999) for a local process of scientizing Chinese herbal drugs, and Bridie J. Andrews, "The Making of Modern Chinese Medicine, 1895–1937," (Dissertation, University of Cambridge, 1996), for the impact of the Japanese modernization of Chinese medicine in China.
[68] See Chen Shengkun, "Zhongguo kexueshe shengwu yanjiusuo zhi yanjiu" (M.A. thesis, National Taiwan Normal University, 198?), Chap. 3 and 4.
[69] See Ma Qiusha's, p. 222, and Zeng Zhaolun's "Ershinian lai zhongguo huaxue zhi jinzhan" (The development of Chinese chemistry in the past twenty years) in *Zhongguo kexue ershinian* (Chinese science—the past twenty years) (Shanghai: the Science Society, 1937), pp. 87–90. Even the Japanese research institute in Shanghai had launched such research in the 1930's.
[70] Grant's report was collected in *Methods and Problems of Medical Education*, series 14 (New York: the Rockefeller Foundation, 1929), pp. 109–145. John Grant was also the Co-director of the Rockefeller Foundation's progressive rural China program. See Mary Bullock's chapter VI: "John Grant, Medical Bolshevik" for his career in China.
[71] See Chiang Yun-chen's Chapters 8 and 9, and Ma Qiusha's chapter 6: "The PUMC Dilemma, the Rockefeller Foundation's Ideal Versus Reality," for a detailed treatment of the reorganization and policy change, and internal struggles among different opinions in the Foundation.
[72] Mary Bullock, pp. 190–191. For the works and roles of PUMC and the Rockefeller Foundation after the outbreak of second Sino-Japanese war, see the whole chapter "The Challenge of War."
[73] For analyses of the colonial viewpoint by the Rockefeller enterprise, see Marcos Cueto ed., *Missionaries of Science: the Rockefeller Foundation and Latin America* (Bloomington: Indiana University Press, 1994), and Soma Hewa's *Colonialism, Tropical Disease and Imperial Medicine: Rockefeller Philanthropy in Sri Lanka* (Lanham: University Press of America, 1995). See the appendixes for the Rockefelller Foundation's expenditure in China and other countries.
[74] Suzuki Zenji, *Nihon no yûseigaku: sono shishô to undô no kiseki* (Japanese eugenics: its ideas and locus of movement) (Tokyo: Sankyo, 1983), pp. 67–68. Tôyama submitted this experiment report in 1906.
[75] Ibid., pp. 66–69. I was informed that actually Mendel himself had done theoretical biometric study, even before Galton's research, in the second half of his original paper, which was oddly overlooked by biologists and historians of the time and later periods.
[76] Ibid., pp. 69–70.
[77] See Ernst Mayr, *The Growth of Biological Thought*, pp. 714–738, for a brief account of Mendel's findings, contributions and unsolved problems.
[78] *Nihon no yûseigaku*, pp. 70–71.

[79] *Seimei ron* (Tokyo: Rakuyôdô, 1915, pp. 320–356. Mendelism was used here as hereditary principles to predicate that genetic diseases would invariably manifest themselves in later generations even though they remained unexpressed temporally.

[80] *Nihon no yûseigaku*, p. 103.

[81] Suzuki Zenji, "Geneticists and the Eugenics Movement in Japan" and "Nihon niokeru jinrui idengaku to yûseigaku no kanren: tokuni Komai Taku, Kawakami Riichi, Furuhata Tanemoto no baai" (Human genetics and eugenics in Japan: a case study of three geneticists) in *Kagakushi kenkyû* (Journal of history of science of Japan), no. 101, 1972, pp. 10–16. Works on human genetics did not come out until 1926 in *Idengaku zasshi* (the Journal of genetics). From 1926 to 1930 four articles were produced there, 1931–35, 10 (4.6%), 1936–40, 29 (8.7%), 1941–45, 11 (4.8%). The geneticists' hidden agenda was pinpointed in Nakamura Teiri's "Marxism and Biology in Japan" in *Science and Society in Modern Japan*, ed. by Nakayama, Swain, Yagi, p. 259.

[82] "Nihon niokeru jinrui idengaku to yûseigaku no kanren," pp. 10–16.

[83] Op. cit., pp. 35–43.

[84] Chen Shengkun, "Zhongguo shengwuxue shiyanpai yu diaocha pai zhi lunzhan" (The debate of Chinese biology: Experiment clique versus Survey clique) in *Kexue Yuekan* (Science Monthly), vol. 15, no. 8, 1984, pp. 603–609. This article was developed from his aforementioned M.A. thesis.

[85] Ibid.

[86] Ibid.

[87] Any scientist in that time, if he or she got the chance, would recount how they went about scientific practice. A few such scientists were Wong Wenhao, Ren Hongjun, Bing Zhi, Zheng Taipu, and Hu Mingfu. The members of the Science Society had produced abundant treatments of scientific methodology, which we can sum up by excerpting D. W. Y. Kwok's concise summary: 1) the need for observation, hypothesis, experimentation, and the return to observation—the empirical principle; 2) to achieve exactitude in measurement such a method must employ quantitative means—the quantitative principle; 3) to deal with causal relations and use abstractions to represent them, the method must locate meaningful recurrences of behavior and then formulate general laws or equations which describe and explain such behavior—the mechanical principle; 4) a general assumption of all scientists inherent in their concept of research—the principle of progress by means of science. See Kwok's *Scientism in Chinese Thought 1900–1950* (New Haven: Yale University Press, 1965), p. 21.

[88] Op. Cit., no. 6, pp. 629–636.

[89] *Yichuanxue qianshuo* (An introduction to genetics) (Shanghai: Zhonghua shuju, 1926), Chapter VIII, pp.139–151. Chen was also the author of *Jinhualun qianshuo* (An introduction to evolutionary theory, 1924).

[90] For instance, his articles "Hezhongguo jueyulu zhi xianzhuang" (The current condition of sterilization in the United States,), 1923, "Yousheng gailun" (An introduction to eugenics), 1924, "Ershinian lai shijie zhi yousheng yundong" (The eugenics movement in the past twenty years), 1925, and "Jindai zhongzu zhuyi shilue" (A brief history of modern racial studies,), 1925, were written in New York while he was studying with Charles Davenport at Columbia University. The above

articles in the volume *Yousheng gailun* were published in 1928 and were later compiled by Pan's daughter Pan Naimu into a Collection series, *Pan Guangdan wenji* (Beijing: Beijing daxue, 1993). *Yousheng gailun* appears in the first volume of the Collection.

[91] Ibid. His three articles of 1927 were: 1) "Du 'du zhongguo zhi yousheng wenti'" (A response to the critique of 'the Chinese eugenic problem'), a response to Zhou Jianren's critique of his "the Chinese eugenic problem"; 2) "'Xinwenhua' yu jiakexue: bo Zhang Jingsheng" ('New Culture' and pseudoscience: repulsing Zhang Jingsheng), a critique of Zhang Jingsheng's false presentation of sexual techniques and their connection to eugenics in the name of science; 3) "Wulin youlan yu renwen dilixue" (A visit to Wulin and its human geography), an association of Chinese historical figures and their relevancy to the specific locale Wulin).

[92] See his work entitled *Zhongguo zhi jiating wenti* (The problem of the Chinese family), published in 1928 and now in his Collection, vol. I, pp. 67–241. In order to prevent a digression, here, I postpone going into his analysis further. Part of his analysis will be discussed in the following chapters.

[93] Pan's Collection, vol. II, pp. 440–465.

[94] See Daniel J. Kevles's *In the Name of Eugenics*, p. 45.

[95] See Ding's article "Zhesixue yu pudie" (Eugenics and genealogy) in *Gaizao* (Reconstruction), 1919, vol., 3, nos., 4, 5, 6. Ding did make a critique of the shortcomings of the recording system of Chinese family genealogies, in that matrilineal information, female offspring, early deaths and physical characteristics were seldom recorded. He also explained how one could best benefit from these materials if one looked at only the reliable elements such as dates of birth and death, number of progeny, official ranks or academic degree, and number of wives and concubines who reproduced heir and successor.

[96] Suzuki, *Nihon no yûseigaku*, pp. 104–105.

[97] Ibid., p. 146.

3

LAMARCKISM VERSUS MENDELISM—THE POLITICS OF BODY AND HEREDITY

ALTHOUGH JAPANESE GENETICS STUDY WAS BASED ON MENDELISM, JAPANESE geneticists were cautious in applying it to human genetics. In China, Chinese geneticists were even more cautious in using Mendelism and merely appropriated it for animal and plant breeding. In addition to the disciplining structure of scientific processes and the culture of local science which I discuss in chapter two, an implied reasoning for this cautiousness was that, according to Mendelism, characters were visualized as units transmitted independent of growth, and therefore newly acquired somatic modifications could not be transmitted to the succeeding generations. If Mendelism was applied thoroughly to human beings, there would never be a possibility of racial improvement without preventing the procreation of the unfit, according to the Mendelian law of segregation.

In this case, Japanese and Chinese, seen as "inferior" in the eyes of Social-Darwinians in the West, would not become "fit" unless they kept interbreeding for centuries with the claimed "superior" white races. Ideally, if they were able to implement white-yellow intermarriage successfully and continued breeding extensively without crossing back, what sense would it make to constitute a distinctive "Japanese" or "Chinese" race as opposed to the other races after a couple of generations? Becoming fit while remaining "Japanese" or "Chinese" was a challenging dilemma—a parallel to the modernization dilemma of "how to become modern while not losing an identity"—which the Japanese and Chinese eugenicists needed to wrestle with. This dilemma was implanted when the ideas of racial improvement were initially introduced into the East, as exemplified by Katô Hiroyuki's critique of Takahashi Yoshio's proposition of the yellow-white intermarriage. The eugenicists had to redefine a racial hierarchy and situate Japanese and Chinese genetic traits and determine their fit and unfit characters within the domain of national body before they could bring up any specific eugenic solutions.

In this chapter, I argue that Lamarckian ideas did play an important role in different ways in the eugenics movements in Japan and China, since Lamarckism offered a more humane alternative to the depressing aspect of Mendelism and provided a foundation for the wishful belief that evolution was necessarily progressive. Ideologically, the idea of Lamarck's time that somatic modifications acquired from an organism's development of particular habits would pass on to the offspring under appropriate conditions, sustained a prospect of improving the human race through social reform and suggested that an improved environment would produce better people. As in the Soviet Union, but in a different way, the popularity of Lamarckism in China and Japan exemplified the truism that any intrusion of ideology into science fosters wishful thinking. Because none of the critics and advocates discussed in the following sections were Lamarckian biologists in a strict sense, this chapter considers Lamarckism as a series of ideas emphasizing environmental factors and thereby could be applied as a social-control project under Jean-Baptiste Lamarck's tenet of inheritance of acquired characters.[1]

My consideration of Lamarckism is inspired by and grounded upon Nancy L. Stepan's studies of the appropriation of Lamarck in the eugenic projects in Latin America and especially Brazil. Stepan shows that whereas a Lamarckian-style genetics could accommodate the language of modern Mendelism and accept Mendelian laws of inheritance, it however left a space for the idea that somehow an influence from the social milieu could permanently modify the germ plasm. Culturally and intellectually tied to the French source, many Latin American doctors saw Lamarckian ideas such as homiculture and puericulture (the scientific cultivation of children involving the knowledge of prenatal care, medicine and obstetrics), as extending principles of public health into the special sphere of heredity in reproduction. "Eugenics thus became linked to obstetrics, population policies, infant welfare, and made common cause with campaigns against alcoholism, tuberculosis and venereal disease."[2] In early 1920s Brazil, eugenics was structurally and scientifically congruent with sanitation science and was perceived as simply a new branch of hygiene. The Brazilian eugenicists furthermore insisted that "to sanitize is to eugenize."[3] Obviously, the meaning of eugenics in the Brazilian context was inseparable from public hygiene and other related environmental programs, and the same merger of signification also occurred in the Chinese and Japanese contexts.

This chapter is divided into four sections. The first section examines the symptoms in the "national character" representation and the eugenicists' presentation of the need for eugenic measures. The second section lays out the various diagnoses and remedies of public-health problems presented by the public hygienists, socialists, sociologists and eugenicists, and their subsequent social debates. The third section concludes with the disparate outcomes from these social debates: whereas a consensus on a comprehensive eugenic bill was achieved by both hygienists and eugenicists in the Japanese context, the Chinese eugenic positioning on heredity was vilified by a series of vigorous attacks from both the rightists and leftists who emphasized nutrition, education, clean surroundings, epidemiological work, medical care and especially prenatal care to better the human race. As a result, the Chinese eugenic propositions were predominantly Lamarckian ones as defined above.

Readers will find this chapter more or less similar to conventional parallel comparativism. The reason for this inevitable style of writing is that Japanese and Chinese eugenics as developed in the 1920s and 1930s focused on many local problems, and the eugenicists tried to raise public awareness and involve the State into their social-reform programs by provoking cultural criticism and public debates, from which each citizen would be able to connect his or her individual body/spirit to the national body/spirit. In the debates, the viewpoint called Lamarckian often referred merely to environmental improvement, sometimes with no implication that environmental influences were transmitted genetically. Hence, I follow the debaters' nonstandard usage for consistency.

SYMPTOMS: NATIONAL CHARACTER AND THE EUGENICISTS' PRESENTATION

The discourse on national character which arose in Europe in the 18th and 19th centuries not only portrayed the uniqueness of one's country, but also justified Western imperialist expansion by taking the representational advantage of subjecting power over the depicted object. Through translational practice, this national-character discourse was imported to China and Japan at the end of 19th century. Intriguingly enough, whereas the Chinese deployed this discourse of self-knowledge into self-criticism to reconfirm unintentionally what had been represented by the West and later by Japan, the Japanese thoroughly embraced both sides of subject/object or self/other positionings. On the one hand , they overturned how Japan was represented in the production of Western Orientalism. On the other hand, they took the same type of advantage over other nations like China (including Taiwan), Korea and South Asia, presenting them as depicted objects, while taking Japan itself as a unique entity.

This twofold imaging of the Japanese national-character discourse stemmed from the framework of the mutually exclusive binaries *chûgoku/iteki*, civilization/barbarian, and inner (*nai*)/outer (*gai*), as it had since the Tokugawa period, when China functioned as the "Other" to help define the Japanese "self."[4] Later, from the Meiji era on, the representation of a monolithic West took over the previous function of China to help mirror for the Japanese their own essence of indigenous culture. On the other hand, China, through the *Tôyôshi* (Japan's Orient) apparatus,[5] was represented as an Orientalist "other" in the shaping of Japanese "self" from the compounding of various binary parts such as East/West, spirit/material, art/science, traditional/modern, past/future into a composite synthesis which was believed to be a realization of the best of East and West by 1920s cosmopolitanist thinkers. Later, in the trend of culturalism of the 1930s, as Tetsuo Najita and Harry Harootunian point out, "[t]he recognition of having achieved this unprecedented synthesis validated the subsequent belief that Japan was uniquely qualified to assume leadership in Asia."[6] If Japan had reached new levels of achievement that no single Western nation ever had, then how did the Japanese eugenicists present their case that there was still room for racial improvement?

In the decade of the 1920s, the eugenicists' presentation of the need for eugenic measures in Japan was initially derived from a series of political and economic concerns. For instance, Tanaka Yoshimaro (1884–1972), the director of the Japanese Genetics Association, wrote in 1924,

> Germany was once one of the world's super-powers before World War I, but now becomes poor and bankrupt. Like Germany, Japan, formerly strong and rich, now suffers economic recession after the Great Earthquake of 1923. The depreciation of the Japanese currency is so devastating that Japan is treated by other countries as a second-rate nation. However, in view of the fact that the fundamental qualities of the Japanese people are still intact and prevail in the new generations, it is an extremely pressing need to devise a fundamental measure to safeguard the excellent qualities of the Japanese race in the face of future political and economic upheavals.[7]

In 1925, he contested the Americans' eugenic claim that Japanese immigrants would debase the American racial quality. He saw that this anti-Japanese sentiment in the Immigration Act was rather driven by political and economic interests opposed to the Asian minority. He defended the view that the fundamental quality of the Japanese people was never inferior to that of the Americans. Even though generally the Japanese physical constitution and facial appearance could not compete with those of Americans, these two features were less relevant in the fundamental concerns of eugenics. The nature of the immigration problem was rather seen as individual; in other words, Tanaka redirected the point of the question to be what kind of Japanese individuals immigrated into the United States. Before being able to answer this question, Tanaka solicited support for the establishment of a genetic research institute in order to conduct a thorough investigation of such an important issue.[8]

Another central figure of the eugenics movement was Ikeda Shigenori (1892–1966). A journalist who later studied Nazi medicine in Germany and became a full-time devoted movement leader, Ikeda proposed in 1928 that the goals for the movement were to promote the Japanese race as in the first rank of the world, and to safeguard its leadership in the human culture. In order to achieve these goals, not only should the knowledge of eugenics be popularized, but a social reform should also be implemented to improve the mental and physical constitution of the Japanese people, so as to raise the spirit of self-government and enforce the disciplining project of citizenship. Ikeda perceived that the gist of the eugenics movement was based on the five guidelines expressed in the Imperial Rescript on Education (*Kyôiku ni kansuru chokugo*) promulgated in 1890.[9] according to Carol Gluck's analysis,

> [The Rescript on Education] consisted of a Mito school beginning, which linked kokutai (national polity, a personified embodiment in the emperor system) with loyalty and filiality; a Confucian center, which enumerated the virtues that inform private and public human relationships, adding a modern injunction to civil obedience and national sacrifice; and an imperial ending, which made the emperor the source of a morality that was said to be both indigenous and universal at the same time.[10]

Identifying with the nationalist trajectory set out by the Rescript on Education, Ikeda overcame the eugenicists' dilemma of how to remain "Japanese" in the process of racial improvement. Moreover, Ikeda saw the eugenics movement as the most effective means to preserve the national traits (*minzokusei*) and national spirit (*minzokukon*).[11]

In the 1930s, we can see a rhetorical emphasis in their presentation showing much confidence in the "Japanese race." Using the metaphor of "jade burnishing,"[12] Nagai Hisomu, the chairman of the Japanese Racial Hygiene Association, advocated the need for eugenics for the Japanese race. He unraveled the dilemma of Japaneseness and racial improvement with the following statement:

> As much as environment acts on the species, education is surely important to the human being. Nonetheless, it is absurd to think that one can burnish and make shine a tile in order to turn it into a jade. This is the poignant truth we learn from modern genetics. It is man's body and life, offspring, family, country that constitute what man is. It is Galton's hand to wake us up about the fundamental problem for man's future. . . . Recently, there is a national appeal for arousing the Japanese Spirit and the glory of national polity (*kokutai*). If the nation of Japan were considered in a concrete sense of "the Japanese people," in order to devote our loyalty to the nation and filial piety to the ancestors, we should sustain and protect this national body which

> was shaped and formed through a series of cultural assimilation processes and military campaigns for two thousand and six hundred years. No other races and nations have reached such a glorious historical achievement and maintained such longevity. Our veins convey valuable hereditary elements descended from our ancestors that we have to examine, protect and extend with the knowledge of Racial Hygiene (minzoku eiseigaku). In order to awaken a distinct sense of Japaneseness in the nation of Japan, from now on, nobody can refuse the task of reviving the Japanese Spirit and the national body. Aiming at this particular goal, we contributed our individual efforts to the establishment of the Japanese Racial Hygiene Association [in 1930]. We now need everyone's pledge to this campaign.[13]

Inspired by his predecessors, scholars like Ogyû Sorai (1666–1728), Kamo no Mabuchi (1697–1769), Motoori Norinaga (1730–1801)and Hirata Atsutane (1776–1843), who did not discard the valuable essential qualities of Japanese culture in the face of cultural domination of Confucianism from China, Nagai expounded the goal of biology and other natural sciences as a support for the unique Japanese national polity. Like Ikeda, Nagai's position on the application of eugenics not only transcended the question of how to retain "Japaneseness" in the eugenic project, but also reversed the racial hierarchy imposed by Western Social Darwinism through a celebration of the glorious historical continuity that no other race or culture had ever achieved.[14]

Such a positing of the supremacy of "the Japanese race" was not plausible in the Chinese context, since the Chinese national character was depicted as decadent and degenerated by May Fourth Era intellectuals such as Lu Xun and Yu Dafu.[15] If the relationship between eugenics and the national character was figuratively like a vitamin supplement given to a healthy person in the Japanese eugenicists' presentation, then in the Chinese case it would be rather like remedies given to a "sick man of the East (*dongya bingfu*)."[16]

In 1929 and 1933–35, Pan Guangdan translated and published respectively Ellsworth Huntington's four chapters on the Chinese character in *The Character of Races* and Arthur H. Smith's fifteen chapters in *Chinese Characteristics*. Pan later included both works in his book *Minzu texing yu minzu weisheng* (1937, National character and racial hygiene). Different from the previous Chinese rendition of Arthur Smith's work from the Japanese translation, Pan selected fifteen of the twenty-seven chapters of the entire book and reorganized them in his own sequence of physiological and psychological, economic, and social categories. After he juxtaposed the translation of Smith's and Huntington's works, Pan also included three chapters of his own writing. The introductory chapter was a critique of contemporary discourse on the themes of "nation (*minzu*)," "national character (*minzu texing*)," and "national revival (*minzu fuxing*)." The fourth chapter was on the symptoms of national diseases (*minzu bingxiang*), while the fifth chapter discussed the prospect of racial hygiene. In this reorganized work, Pan suggested to his Chinese readers that through the mirrored image reflected from these two foreign experts' writing one could come to grips with the characteristic symptoms of the Chinese people.[17]

The symptoms of the Chinese national character, Pan observed, can be divided into five aspects—cultural, social, economic, physical and the will for survival—which were based on their contrast with an ideal healthy nation that has rejuvenating cultural creativity, facile adaptability, robust physical constitution and invincible will for survival. Culturally, Chinese creativity had been vastly diminishing, especially seen in the poor performance on Ministry of Education general examinations in mathematics, foreign language (English), physics and chemistry at the primary and secondary school level. Unlike Japan, in the domain of overseas education there were pitifully few Chinese students attracted to the fields of science and technology. This deficiency of interest, according to Pan, was not necessarily due to a lack of scientific education, but was perhaps a demonstration of inherent incapability. At this point of time, China had undergone for three decades or so various social reforms promoted by different generations of returned students who received their overseas education from different countries. However, none of the reforms could synthesize the best parts of these manifold fashionable ideas, and come forth with suitable policy. Pan commented that this failure was another indication of inadequate creativity.

Socially, Pan continued, there had been no well-organized structure in the society to hold up the various social forces in the face of the political disintegration of recent decades. Bureaucratic inefficiency was a prevailing syndrome, from the top level of the national government, all the way down to the administrative organ in small regional schools. Even the overseas Chinese had the same problem in that the Japanese colonizers in Southeast Asia easily outcompeted Chinese enterprises in their "Marching Southward" movement. China did have strong social organizations such as kinship, guild, and old-fashioned village and secret societies, whose networks were based on their particular interests, but none of these could transform into an impartial political organization. In view of incessant internal sociopolitical chaos, China was very vulnerable to foreign invasion.

Economically, Chinese industrialization was far underdeveloped. Lack of creativity and of organization were two of the main reasons for the inadequate means of production. In addition, the most weighty shortcoming, as mentioned in both Arthur H. Smith's and Ellsworth Huntington's works, was the "selfishness" rooted in the Chinese national character. Selfishness proliferated official corruption, and promoted the two extreme aberrations of avarice and profligacy, which contributed to an uneven distribution of social wealth. Although Chinese people were diligent hard-workers, they were not able to increase productivity without efficient management.

In terms of physical constitution and vitality, Chinese, like Jews, were said to have a high capacity to endure in or evade any unfavorable environment. For instance, Chinese emigrants were all over world, surviving from cold to tropical zones. From Pan's writing of various historical migrations from North to South China, his denotation of "Chinese" was certainly Han Chinese. Despite their high death rate, Chinese could also resist various kinds of diseases such as yellow fever, malaria and tuberculosis, more so than other people who might have been devastated by diseases turned into disastrous epidemics. From an unnamed source, Pan made another example of the Chinese ability to resist syphilis, in that Chinese did not suffer as much as the Whites the attack on the nervous system in the tertiary stage. The Chinese ability to endure, like sticky candy, was said to be so flexible that most Chinese people were seemingly patient and indifferent to their suffering from harsh surroundings. Some intellectuals characterized this endurance and contentment with temporary ease as survival skills, and considered them positive characteristics of the Chinese infantry. However, in the milieu of ruthless international competition, Pan did not take them as an evidence of "the survival of the fittest." Pan believed that there was plenty of room to improve the Chinese physique, not necessarily through external physical education and athletics movement, but rather fundamentally from a genetic buildup.

Pan criticized that, although Chinese had strenuous will as seen from the above example of frontier migration and settlement, they were only selfishly concerned with their individual or lineage propagation, without thinking about the nation as a group unit for survival. In addition, Chinese did not have any desire to glorify their national life, celebrate heroic spirit or venture into the untrodden world. After a series of foreign invasions, opium, civil wars, anti-superstition campaigns, and custom-reform campaigns, the Chinese people had learned how to manage barely to survive. The Chinese national spirit, Pan judged, was exhausted and decaying.[18]

Pan's representation of symptoms in the Chinese national character resonated with the May Fourth critique on cultural decadence and racial degeneration of the Chinese national character. The May Fourth intellectuals tended to present the Chinese civilization and physique as overripe and declining. Chen Duxiu, using the metaphor of new and old cells for the youth and old, saw in China that the new and lively cells had been infected by the "virus" of the old and decaying ones. He said, "When I look at our youth, five out of ten are young in age but senile in body, nay, nine out of ten are young in age and body, but senile in their brain nerves."[19] In the same fashion, Lu Xun criticized, "Chinese males and females tend to age ahead of their time; before they reach twenty, they are already senescent."[20] Thanks to this degeneration discourse, eugenic ideas became a possible source of inspiration for national salvation and started to gain increased attention from the public. For Pan Guangdan, eugenics would be the key to regenerate the race and nation. For Lu Xun and Chen Duxiu, the only outlet for rejuvenating the old decaying civilization was to revive its "instinct," or search among beastly barbarians for new blood. The historian Gu Jiegang (1893–1980) furthered the gist of this representation and stated that the corruption of Chinese civilization needed to be rescued by the fresh energy of its primitive and virile neighbors. Chinese history had shown that in the South-North division epoch (420–581), the Five Dynasties epoch (907–960), and the Yuan (1206–1341) and Qing (1644–1911), China had received infusions of fresh blood.[21]

However, there was a nuance of difference between Pan Guangdan's and the May Fourth representations. Whereas Lu Xun and Chen Duxiu saw feminization and senility in the Chinese civilization and physique, Pan was not gloomily pessimistic and proposed that the Chinese nation, instead of being aged, was still in its youth. Using the anthropological study of Griffith Taylor's *Environment and Race* (1927), Pan maintained that the Latin peoples were the oldest races, with the Teutons second and the Slavs third. The Chinese people were historically mixed by various races, some old and some young, but compared with the Latins as old-aged, and the Teutons and the Slavs as middle-aged, Chinese were still youthful. Pan explained that, viewing the Chinese people as a sickly youth struggling to grow up into a strong mature man, he gave the eugenics journal he established in April, 1932—right after *Yousheng Yuekan* (Eugenics Monthly) was discontinued—the name *"Huanian,"* an abbreviation from the phrase of helping the Chinese nation to reach maturity.

On the one hand he affirmed that there was enormous stamina (*Yuanqi*, a term having a genetic or prenatal connotations) in the Chinese nation. On the other hand, he felt obliged to admit that China had received a corrupting influence from certain historical negative selection in its childhood, and that its genetic constitution was now suffering various pathological symptoms. In order to restore it to its original health, Pan advocated that a social device of cultural and natural selection based on eugenic principles should be implemented as soon as possible. If Japan could become a great nation like Turkey, Italy and Germany, Pan believed that surely China could do it as well.

Pan's nationalistic expression of Chinese national character, and his presentation of eugenic proposals in wrestling with the dilemma between the Social Darwinian racial hierarchy and national identity, were recast and shaped by the materiality and intellectual discourses of that period. As Pan's best friend Li Jinghan remarked in the preface to Pan's book, one can improve oneself only through self-critique and self-knowledge. Pan had to strategically present cultural critiques in order to promote eugenics as a hopeful tool to help China survive in the relentless competition, whereas the Japanese eugenicists had transcended the dilemma and reversed the racial hierarchy before the implementation of eugenics.

Interestingly, none of the Japanese and Chinese eugenic presentations of symptoms belonged to the categories of genetic traits, but were rather phenotypic traits developed from cycles of cultural and environmental influences. Hence, they could unclog the impasse between national identity and inherent inferiority. Since they had set the focus of cultural critiques in their presentation of symptoms, we should not be surprised that the diagnoses and prescribed remedies set forth in the debates merged both Lamarckian (hygienic) and Mendelian (eugenic) approaches.

DIAGNOSIS AND REMEDIES IN EUGENIC PROPOSALS: LAMARCKIAN OR MENDELIAN?

Despite the wishful representation of the supremacy of the "Japanese race" among the eugenicists and other intellectuals, Japan, like other countries, had to face the difficult reality of public-health problems. In addition to various virulent endemic diseases, there were other menacing diseases besetting the Japanese state. Tuberculosis turned into an epidemic in 1880–1900 and declined only after the end of World War II when the antibiotic streptomycin and other drugs were imported and the vaccination of BCG was implemented. During this interval, the average tuberculosis mortality in Japan was roughly 200 per 100,000, two or three times higher than in European countries such as Britain, Germany, Holland and Belgium.[22] In terms of leprosy, according to Shimizu's statistics, in 1900 a total of 30,359 Japanese lepers were being treated. In 1906 it decreased to 23,819, and it was gradually reduced to 11,326 in 1940. Meanwhile, around 1930 there were only one thousand lepers in the United States.[23] As for venereal diseases, in 1920 the frequency among youth waiting for military conscription had reached the total of 11,326, of which 70% were contracted from prostitution. In the 1933 national health survey of unlicensed prostitutes and women in related business, among the total of 1,915,318 examined, 64,422 had venereal diseases.[24] The average number of reported cases of mental illness, including mental retardation, between 1926 and 1930 was 66,699 and the annual increase was about three thousand sufferers. (Japan's population then was about 73 million.)

The picture of public health in China was even more melancholy. Tuberculosis was one of the fatal diseases. According to John B. Grant at the Peking Union Medical College, although there were no statistics from the government, local records showed that in Beijing around 1927–28, the mortality of tuberculosis was over 500 per 100,000, probably equivalent to the death rate prevailing a century earlier in Europe.[25] One can imagine only that the scenario was worse in areas outside Beijing, the capital, with a lamentable lack of adequate sanatoria and facilities for proper treatment. As for lepers, it was estimated by the Chinese Mission to Lepers that there were about one million sufferers from this disease, approximately one-half of the world's lepers. Venereal diseases were widespread in China but there were no national statistics on these diseases. However, one glimpses the tip of iceberg by comparing the extent of venereal disease among Chinese soldiers with those of other nations. For instance, the rate of syphilis was 219.3 per thousand on an average, and more than 350 in Suzhou and Beijing, while the British army rate for all venereal diseases was 41.4 and the U. S. army was 49.9 per thousand in 1926.[26] For mental disorders, there were no records from the central government and it was not considered by the ordinary hospitals as a category for treatment. The extent of mental diseases was estimated as about one in 400, a rate similar to that of Japan.[27]

Facing such pathological phenomena in both Japan and China, hygienists, sociologists, cultural theorists, educators, and philosophers provided their various diagnoses and thereby remedies different from those of the eugenicists. In order to encompass the multi-approaches within the Lamarckian umbrella, it is important to specify their numerous statements on this particular issue and in the form of debates. In the following discussion, one can find different degrees of synthesis between Lamarckian and Mendelian views in both China and Japan.

A. CHINA

As early as the establishment of the National Medical Association (later changed to Chinese Medical Association) in 1915, eugenic ideas were incorporated into the knowledge of social hygiene promoted by the Association's journal *National Medical Journal of China*. A medical excerpt "Shehui weisheng lun" (Treatise on social hygiene) of 1916, translated from the Japanese journal *Chuo koron* without acknowledging the author and its detailed source, introduced important measures of Japanese public hygiene, such as population surveys, in relation to such matters as the balanced distribution of men and women, labor and leisure, nutrition, home hygiene, public bath, procreation, motherhood, breastfeeding, school hygiene and social welfare. In the section on procreation, it warned against the elements of tuberculosis, alcoholism, venereal diseases, and mental diseases as negative choices for marriage and reproduction, by giving an example of their potential for procreating hideous offspring prone to crimes. Although such degenerate elements of offspring were not yet found in Japanese nationals, it was easy to observe many unqualified physiques in the conscript examination. As the eugenics bills were implemented in the United States to preclude criminals and the feebleminded from marriage and from increasing the problems in social hygiene, both Japan and China should watch closely their development. A related social problem was in birth-control methods in

China, for which it was thought that the long-standing practices of abortion and infanticide needed to be curtailed.[28]

In another issue of the same Journal, we see in the column of public discussion an interesting juxtaposition of two articles by Wu Liande (1879–1960, head of the North Manchurian Plague Prevention Service and later a director of the National Quarantine Service) and Yu Fengbin, two of the founding members of the Journal. In Wu's article, he cast a confident outlook for the future health of the Chinese nation and race. He said that if the preventive plans of anti-epidemic, mass education of public hygiene and improvement of medical facilities were properly adopted, the health of the nation would be gradually recovered and the state would become strong and competent. For Wu, plagues, infectious diseases, famine, civil wars, and insanitation were the determining causes for fatal weakness of the nation and race. In contrast, Yu's article on the causal relationship of marriage and family hygiene emphasized genetic defectiveness and its danger of reproducing a sickly family, and thereby an impoverished society and long-suffering nation. Yu praised the American eugenic implementation in the marriage law and tried to remind his fellow countrymen of the importance of spouse-selection. He evoked the ancient exhortation that inbreeding by kin marriage often procreates problematical offspring that are either physically or mentally handicapped. He also warned against the practice of early marriage in China as violating the eugenic principle. According to his view of Mendelian laws, it was never accidental that people were born of robust or ailing physiques; therefore, it was important to investigate family genealogy before marriage. Moreover, using the principle of the distinction of germ cells and body cells, attributed to August Weismann, Yu asserted that the parents' talents and behavioral traits would not pass down through generations, and that therefore the effects of deformed feet from foot-binding or muscle buildup would not persist in the next generation. However, diseases like syphilis, tuberculosis, alcoholism and narcotism were thought to poison the body cells and subsequently damage the germ cells of the developing fetus. Yu urged earnestly that if one cared enough for family prosperity and health, one should eradicate evil and unhygienic habits and select infallibly one's spouse.[29]

Such incorporation of eugenics into marriage and public hygiene was also advocated by Hu Xuanming (1887–?, M.D. Johns Hopkins University), who was another founding member of this journal and who was in charge of the municipal health department in Canton during the 1920s, where his work on public health provided a model for other cities after 1927.[30] His half-translation-and-half-summary article "Hunyin zhesixue" (Eugenics in Marriage, no source indication of original authorship),[31] published across seven issues in 1919–21, was sponsored by the Joint Council on Public Health Education. The Joint Council was formed by the National Medical Association and the China Medical Missionary Association in cooperation with the Young Men's Christian Association (Y.M.C.A.) in 1916, for which Hu served as secretary in 1920–21. The Joint Council invited Hu to seek out various studies and translate them swiftly into Chinese in order to spread the knowledge of survival in the era of natural selection, since eugenic bills were already implemented in some American states and few Chinese had heard about this development. The Joint Council urged readers to envision their individual marriage choice as a national problem of population, since the majority of the 400 million Chinese were weak and feeble and would be eliminated in the world's competition if eugenic measures in marriage were not promulgated.

Although this article was a translation, Hu sometimes interjected his own opinion. Hu first stated the significance and the scope of eugenics, which goes beyond the conventional understanding of prenatal care (*taijiao*). Instead of emphasizing pregnant women's lifestyle and behavior, eugenics underscores the appropriate marriage matches in order to reproduce better offspring. Hu considered venereal diseases the most detrimental to progeny and recommended that every young woman should demand that her future fiancé obtain a VD health certificate before accepting his marriage proposal. To Hu, eugenics was the most simple and effective formula to save the society and the nation from financial burden and racial degeneration. He upheld that the eugenic consideration in marriage should be a prevalent commonsensical practice in the society, and that the state should preclude any undesirable elements from marrying as well.

Hu also provided a section on basic genetic principles for mating. It is interesting to point out that since the genetic principles were excerpted from the American experience, Hu did not follow the English original and come up with a specific hierarchical standard of hair, eye color, skin, height, weight, talents, and physical strength for Chinese in their marriage selection. Hu subsequently enumerated a series of genetic diseases to serve as background knowledge for genealogical observation and marriage decision-making.[32]

The pattern of eugenic ideas favored by the hygienists was also followed by Hu Dingan (M.D., Ph.D., University of Berlin, 1926). From 1929–37, Hu advocated the state's reform of Chinese medicine and the scientific transformation of pharmaceutical drugs from traditional medicinal herbs, as well as the elimination of traditional medicinal practice from the national education system.[33] In his medical commentaries, Hu diagnosed the reasons for Chinese weakness. Historically and culturally, the Chinese people overemphasized mental activities and ignored military spirit and physical exercise. This historical tendency was further worsened by the depraved customs of foot-binding and opium addiction, the ignorance of medical knowledge, and the absence of national identity and a national health policy administered by the state. Subsequently, Hu proposed a health-strengthening plan with a step-by-step formula. First, rehabilitate the national defense discipline by pursuing national sports activities, which will reinvigorate the citizens' physical strength and the soldiers' military spirit. Second, improve racial stock with better breeding in both mental and physical aspects in order to match up the average standard of intelligence, character and morality. The eugenic measures, according to Hu, included childbirth improvement, marriage counseling, and the fundamental prevention of venereal diseases, tuberculosis, mental diseases and narcotism. Third, concentrate manpower and intelligence on social construction in order to bring forth various statistical data and enhance creativity. Fourth, replenish the already-depleted national strength, which had suffered from incessant epidemic outbreaks, famines and wars. A national health plan should be implemented together with sports education from the primary level of the school system in order to safeguard the foundation of a healthy nation. Fifth, aim high to become one of the great powers on the model of the Euro-American countries. To achieve such a goal, the first four steps should be followed closely.[34]

From the above standpoint of hygienists, we do not see any particular confrontation between hygienists and eugenicists, since some eugenic ideas were already incorporated into a holistic plan of social hygiene. However, this social unity of hygienic and eugenic ideas was rather formed on the basis of the pragmatic concerns of a comprehensive plan of control and prevention for venereal diseases, tuberculosis and mental diseases, which were easily seen and spread in the crowded, urbanized and industrialized cities. This inclusive approach to social hygiene was also seen in the history of the health campaigns of the Young Men's Christian Association (YMCA, established in 1895 initially) in China. From 1912, the Association formally organized a lecture bureau to satisfy the critical need for education and to feature a campaign of propagating public-health knowledge. In order to convey effectively the importance of national health, the bureau employed visual aids such as cartoons, popular literature, slides, films and exhibits to attract larger general audiences, and mobilized student-volunteers, local doctors and local elite to form an efficient community network in each city.[35] In addition to social service and educational projects, the Association's student department attracted a new generation of educated urban youths who were earnestly searching for ways to strengthen China, and formulated student programs of athletic teams.[36] Interestingly, when Pan Guangdan set up the *Eugenics monthly* (*Yousheng yuekan*) from June, 1931, to March, 1932, and the other aforementioned eugenics journal *Huanian* from April 1932 to 1937, both journals were sponsored by the YMCA in Shanghai.[37]

Nonetheless, this inclusive hygienic campaign by no means conformed to the Mendelian standpoint of the eugenicists, in whose opinion a radical and comprehensive plan of environmental improvement could never assure the betterment of race and nation. For the eugenicists, the euthenic approach of medicine, social hygiene, philanthropy, education and economic reform could merely provide temporarily cosmetic change.[38] Viewed as such, the eugenicists' fundamental position was not only different from those of the hygienists, cultural theorists, socialists and educators, but also provoked a series of debates. From the following study of these debates, which shaped the multilateral meaning of eugenics, we can have a better grasp of the difference between the Mendelian and Lamarkian approaches and its nuances in the Chinese context.

Match One: Pan Guangdan versus Zhou Jianren

Pan Guangdan's 1924 article "Zhongguo zhi yousheng wenti" (The problem of Chinese eugenics) raised a question about the effectiveness of various reforms since the mid 19th century: "Will these reforms, amounting to environmental changes, help the Chinese nation and race survive in the world of competition?" Pan compared and contrasted the Chinese situations before and after Westernization. He argued that the survival and progress of a civilization rely upon the mechanism of cultural selection against the force of natural selection. Pan felt assured that, with so little cultural selection against the force of natural selection, the original constitution of the Chinese germplasm must have been vigorous to survive throughout centuries of natural selection. Before Westernization, in view of poverty, diseases, natural disasters, famine, wars and illiteracy, the law of natural selection governed mostly the evolutionary processes. Given the high mortality rate of infants in China, the survivors were generally the fittest even though there had been cases of survival of the feeble.[39]

Pan further explored eugenic ideas and hereditary principles as the mechanism of cultural selection embedded in the traditional literature and institutions. For instance, in the *Book of Rites* (*Liji*) the chapter on "*Wangzhi*" (King's institution) says that if manifold educational improvement does not make appropriate change, the only solution is to dispel those undesirables out of China proper. The classic *Zuozhuan* (a commentary on the *Spring and Autumn Annals*) records good and evil progeny of various lords and their subsequent behavior. If education and cultural transformation do not exercise their expected effect, the traditional alternative is to relocate and segregate these presumable misbehaving elements.

Pan continued that the so-called "*xiantian* (prenatal)" in literary narrative does not necessarily denote "heredity," because its conventional usage was often tinted with superstitious color—such as a geomantic connection or an astrological correspondence—whenever characters were depicted as extraordinary, talented, good or evil. Thus genius and talents were definitely born by nature and not nurtured with environment and education (*houtian*, postnatal). For traditional institutions, Pan examined the force of cultural selection as manifested in marriage, reproduction, the examination system for public service, and agrarian life. Marriage achieves the purposes of perpetuating family and lineage, serving the elders and individual happiness. In China, the first two purposes were the primary concerns in marriage selection, since elders dominated marriage selection on the basis of arranged marriages of equal standing (*mendang hudui*), in which personal and family integrity, social and economic status, and physical quality were taken into consideration as a whole. In the Chinese civil code, official-scholars were prohibited from marrying prostitutes, theatrical performers and criminals. Using the source *Dictionary of Celebrities* (*Renming dazidian*) to locate the hereditary principle of genius and talents, Pan pinpointed the production of seven great calligraphers in the intermarriage between the renowned Wang and Xie lineages in the periods of Western and Eastern Jin (266–589), and 18 gifted painters from the Wen lineage in the Ming and Qing. Although Pan critiqued the systems of concubinage and chastity, he somewhat dismissed the impact of concubinage due to the non-significant role of sons of concubines in Chinese society, and affirmed the eugenic aspect of chastity for its stabilizing function in the family and society.

With respect to reproduction, most Chinese perpetuated conventional prolific practice and polygamy, and reproduction constantly exceeded the number that could be fed. However, since there was no man-made mechanism to reduce the high mortality, the Chinese population could still maintain its natural balance. Compared with the dysgenic tendency (depopulation of the middle class and overpopulation of pauperism) in Euro-American countries, China had more leeway to implement appropriate institutions to reproduce the best quality of its population.

In Pan's judgment, the examination system of public service was a unique Chinese eugenic instrument and was equivalent to the western Intelligence Quotient test, with which it selected the best elements in the society and served as the ground for arranged mating. As Pan cited, Francis Galton had once praised the Chinese examination system in his *Hereditary Genius* (1869), "I was desirous of obtaining of facts bearing on heredity from China, for there the system of examinations is notoriously strict and far-reaching, and boys of promise are sure to be passed on from step to step, until they have reached the highest level of which they are capable."[40]

In comparison with heavily industrialized urban life, the Chinese rural setting was better for rearing children and maintaining their racial quality. Chinese agrarian dwellers constituted more than 80% of the total population and their quality was relatively robust. Pan suggested that what they needed thus far was proper education.

The above constituted the Chinese eugenic situation before Westernization. In the eyes of cultural iconoclasts like Lu Xun, Pan's affirmative explanation of conventional mechanisms of cultural selection sounded like using the frosting of a new eugenic vocabulary on the cake of an old regime. I think Pan's major strategy was rather eclectic and was not to restore old institutions without any modifications. He was concerned primarily with the question of how to come up with a comprehensive and systematic plan of cultural selection without the risk of dysgenic selection as had occurred in the Euro-American countries. Pan criticized those environmentalists and sociologists who parochially maintained that environmental and social improvement was sufficient to guarantee the progress of social evolution. Pan said that most of these sociologists were limited in their perspective by their Western overseas education experience, where environmentalism was popular because their populations were not crowded and natural resources were abundant. Consequently, when they returned to China, they transplanted what they learned and received in the West without adapting their ideas to local situations.

Pan continued to critique other popular ideas imported from the West, such as individualism, socialism and democracy, for their egalitarian principle and conflict with Chinese familism. An extreme development of individualism would selfishly emphasize individual freedom and happiness over one's social and family obligations. The socialist goal of social security was not necessarily different from that of eugenics; however, its egalitarian principle and environmentalist faith disregarded individual hereditary differences and assumed unrealistically that everyone was in the same position and status in the polity and social organization. Pan exhorted those Chinese socialists to reconsider carefully and clearly these idealistic premises. For Pan, democracy shared the same idealistic premises with socialism. Even though both democracy and socialist reform affirmed equal opportunity of development, they would eventually end up eliminating the best elements in the society and impair racial quality. This phenomenon, Pan warned, had been witnessed in the West, and was developing subsequently in China.

Pan elaborated the dysgenic tendency in the recent social changes derived from Westernization. First, the recent development of medicine and hygiene would reduce the force of natural selection and help the weak survive. Once the death rate of the weak decreased, their reproduction rate would go up; this would force the strong to practice marriage control or birth control and lead to a crisis of degeneration in the population. In a well-developed and sparsely-populated country, this kind of crisis would not occur immediately; however, in China, one could expect this exigency of degeneration before long. Second, the recent promotion of marriage selection in the *Lady's Journal* (*Funü zazhi*) was too individualistic and romantic in orientation (or too westernized in Pan's judgment), without taking the concern of racial quality into consideration. For instance, the principle for seeking the right spouse in the magazine's questionnaire was primarily romantic love. Without free love, many readers responded that they preferred late marriage or to be unmarried. Moreover, only 50% of the readers demanded physical health of their ideal marriage partners and 70% ignored the parameter of hereditary diseases. Among male respondents, 38% preferred their wives to be economically independent and only 33% favored for her a career of housewife and childrearing. Pan lamented that the new youth merely pursued individual happiness and no longer recognized the value of motherhood as a pivotal link for family connection and racial continuity. Third, in addition to the dysgenic tendency of marriage, there were a fair number of intellectuals practicing the fashionable birth control promoted by Margaret Sanger's visit to East Asia. Pan criticized the apparent situation that some celebrities advocated birth control in order to express their idiosyncratic individuality. Pan concluded with a warning that, without following blindly, China should adopt scrupulously the appropriate mechanisms of various kinds of cultural selection from the West.

Pan's basic positions as shown above provoked a different response in 1924 from another eugenicist, Zhou Jianren, who received his bachelor degree at Tokyo Imperial University and who translated Charles Darwin's *On the Origin of Species* in 1947. Zhou first agreed with Pan's conviction as to the importance of eugenic quality, and then questioned the correlation between the infant mortality rate and unfitness, genius and diseases. He elaborated that the high infant mortality rate in lower-income households was due to the social cause that sufficient prenatal and postnatal care for their newborn infants was unaffordable for most proletarian parents, and had nothing to do with their innate quality of fitness. He pointed out from celebrity biographies that many geniuses and talents either suffered from or were susceptible to various diseases. He also used an example of parasites to explain that survival was not necessarily of the best and that the fittest was not naturally the superior. Furthermore, the decline of a race was seldom because of its biological inferiority. For instance, the evidence of destruction of Greek civilization by malaria was insufficient to prove that the Greeks were inferior. The same case can apply to China and its neighboring countries that experienced environmental changes after the advent of the western powers. If they could not adapt themselves deftly to new environments, they would be doomed to perish. Interestingly, however different their views were, both Pan and Zhou disputed the common notion of Chinese inferiority.

Zhou objected that Pan gave too much credit to traditional values such as filial piety, familism, arranged mating of equal standing and the examination system, which merely made married couples more compatible in their personalities and married life, and did not necessarily reproduce great talents. Such artificial man-made selection based on conventional values did not conform entirely to the interests of race or nation, but to those of individual family and lineage. Zhou questioned satirically that if these conventional institutions were eugenic, how come China did not make any progress and was unable to compete with the West? Zhou exhorted his readers that sexual selection was the best reproductive selection, with esthetics as its key element. Observing the natural pattern of courtship in the animal kingdom or primitive races, he suggested that, in addition to wisdom and morality, sexual characteristics of male robustness and female buxomness conformed not only to the standard of natural beauty but also to the needs of racial reproduction.

Zhou said that modern scientific technology improved our eugenics knowledge of hereditary diseases in order to prevent the perpetuation of undesirable elements, and helped develop a better social environment for wellborn members. It was the responsibility of the eugenicists, politicians and legislators to institute a negative measure of either isolation or sterilization on the one hand, to decrease the number of hereditary paupers, criminals, and mental-deficients; and a positive plan to reproduce those who are socially desirable, on the other hand. Before devising a positive plan, the eugenicists and educators should disseminate genetic knowledge and eugenic ideals to the new youth to prepare their sexual selection in the face of the new epoch of free love. Using the example of Karl Pearson as a Marxist, Zhou maintained that socialism was not incompatible with eugenics. Zhou suggested that eugenicists should work hand in hand with socialists to work out a positive social system for the wellborn members.[41] Comparing Zhou's proposal with Pan's eugenic assessment, we are surprised to find that political leftists could be more aggressive than presumable rightists on the issue of eugenics.[42]

In 1927, Pan replied to Zhou's critique on four points: his attitude toward conventional eugenic values, the correlation between fitness and infant mortality rate, class and intelligence, and genius and disease. Pan defended his position that his exploration of eugenic values in traditional filial piety, familism, arranged mating of equal standing and the examination system was not aimed at promoting these old institutions, but at evaluating justly their historical value in their own right. Pan pointed out that among these pre-existing institutions, the examination system was affirmed and praised for its eugenic value by the British eugenicists Francis Galton and F. C. S. Schiller. Schiller, in his *Eugenics and Politics* (1926), compared the short-lived Roman Empire and the longevity of Chinese civilization, and concluded that the examination system was the key survival mechanism of Chinese civilization.

As for the correlation between infant mortality rate and fitness, Pan cited the biometric statistics of Karl Pearson, E. C. Snow, A. G. Bell, A. Ploetz and S. J. Holmes as demonstration of the rigor of natural selection and Zhou's erroneous presentation of social causes for the high infant mortality rate. Infant mortality rate was in inverse proportion to childhood mortality. Frail infants died early, before they turned into young children, and children's vitality and lifespan were inherited from those of parents. The normal tendency of mortality rates for boy infants being higher than that for girl infants indicated that female fitness was superior to that of the male sex. These pieces of evidence, according to Pan, confirmed infant mortality as biologically selective rather than socially determined. Pan criticized Zhou's overemphasis on social determination and his causal connection of a high infant mortality rate with low-income households. Pan advanced his argument that workers who received lower wages did so mostly from the genetic attributes of their lower intelligence and lesser physical strength, and not so much by virtue of social injustice. Like adult mortality, infant mortality resulted from genetic handicap rather than from social problems.

To refute Zhou's argument that many geniuses and talents suffered from or were susceptible to various diseases, Pan cited Lewis M. Terman's *Mental and Physical Traits of A Thousand Gifted Children* (1925) to provide statistics that gifted children tended to be both smart and healthy. They weighed more, learned faster, and reached their puberty earlier. For Pan, it is impossible to achieve the egalitarian goal since there would always be people of genius and talent who would assume leadership in the society. In diametrical opposition, Zhou stood for an egalitarian and proletarian form of eugenics. Zhou later kept on the debate and pointed out in 1931 the dynamics of environment and its dialectical interaction with individual organisms. Citing the research results of W. H. Adolph, who was teaching then at the Yenching University in Beijing, Zhou gave examples such as the following: southern Chinese immigrants to the North tended to reproduce progeny of taller physique than those who remained in the South; Japanese immigrants to Hawaii had a similar tendency of increasing height; Americans who lived in China for a fair amount of time would lower their blood pressure to that of the local Chinese. In other words, Zhou believed in the tremendous impact of diet and climate upon genetic traits, which could be classified as a Lamarckian approach. Accordingly, children's intelligence and physique would be determined by their family background and class status.[43]

MATCH TWO: PAN GUANGDAN VERSUS SUN BENWEN

Sun Benwen (1892–1979) received his doctorate degree from New York University in 1926 and was the director of the Chinese Sociology Society, and chaired the committee on *Sociology Terminology* (*Shehuixue mingci*) as part of the project of registers of scientific terminology sponsored by the National Translation Bureau from 1937. Coming from a training in cultural sociology, and being a student of the sociologist William F. Ogburn, Sun critiqued and denounced in 1929, four fallacies in Pan Guangdan's eugenics. These included the fallacy of applying animal breeding as a model for human rearing, the fallacy of considering cultural influences as biological determination, the fallacy of using the Intelligence Quotient test as a sufficient criterion to distinguish the superior from the inferior, and the fallacy of measuring a person's ability by standards of wealth and power.[44]

Pan responded that eugenics was not an application of animal breeding to the human subject, because he mainly followed the approach of Francis Galton's genealogical studies on hereditary genius. There were biological components in cultural influences that even cultural sociologists could not deny. Even William F. Ogburn had once admitted that novel inventions half depended upon genius. The biological components were difficult to segregate from cultural influences because human activities were conditioned by both elements. As for the I. Q. tests, Pan clarified that it was an invention devised by educational psychologists. Pan admitted that even though it was difficult to design an absolutely objective measurement, there were some indexes for standardization. Hence, the crux for Pan was not the question of whether the I. Q. tests were sustainable, but rather to use all the indexes to ascertain fair accuracy of measurement. Finally, Pan denied that eugenicists judged people's ability by wealth and power. In the Chinese context, poverty was a chronic problem and poor scholars were a common phenomenon. Even though intellectuals were financially deficient, they were highly respected in society.

Sun was not convinced by Pan's defense, and he embarked on another long response. Sun argued that Mendelism was basically the genetic laws acquired from experiments on peas and was used to substantiate the claim of human genetic control. He enumerated various eugenicists such as Edwin G. Conklin and Paul Popenoe for their use of animal analogies for human application. The eugenicists' denial of inheritance of acquired characters was normally supported by their special case studies of unusual families, such as the Juke Family who generated six generations, roughly 1200 progeny of criminals, prostitutes and other social misfits, and the Edwards Family who reproduced 1394 descendants of academia, physicians, lawyers, priests and the like. Sun contested the eugenicists' neglect of environmental degradation. The negative traits identified in the Jukes Family such as idleness, viciousness, lewdness, pauperism, disease, idiocy and criminality were not inheritable, except certain diseases such as insanity and feeblemindedness. Likewise, the highly regarded social positions in the Edwards Family were also not inheritable. Sun commented that since the eugenicists could not provide substantial evidence of heredity of those undesirable or favorable traits, it was extremely arbitrary and thus dangerous to devise and implement eugenics plans of either abstention from or encouragement of reproduction.

The indexes in the I. Q. tests were culturally defined within the system of cultural habits. Sun argued that the tests measured the result of how well sophisticated people have achieved their learning in society. Hence, the acquired ability had to do with the person's training, education and cultural background. The purpose of the I. Q. tests was to understand the person's particular learning system in order to decide the direction of education or career options. However, the I. Q. tests as used by the eugenicists were to determine whether the person was worth an investment in education, meaning that he or she could consequently be deprived of the educational opportunity. For the eugenic criteria of wealth and social status, Sun cited Popenoe's link of poverty and reproduction rate as the eugenicists' class prejudice against the poor.

The debate between Sun and Pan seems to be a contest between William F. Ogburn and Charles Davenport, the American mentors of Sun and Pan respectively, translated into the Chinese context. In 1931, the public debates between Sun and Pan, and Pan and Zhou, were commented on and evaluated by the pen-named Rusong (real name Ren Zhuoxuan, 1896–1990, another pen name Ye Qing), who went to France in 1920 for overseas education sponsored by Li Yuying (1881–1973, Ph.D. in Institut Pasteur, who introduced Pierre Joseph Proudhon, Mikhail A. Bakunin, Peter A. Kropotkin and Jean-Baptiste Lamarck into China), and later became a strategist and theorist of the Nationalist Party.[45] Among the three debaters, Ren saw Pan as a halfway reformer who lingered on and recycled the old values of cultural institutions such as the examination system, arranged mating and familism. Ren continued that Pan's promotion of racial improvement disregarded the value of individual happiness in one's lifetime, especially for women who, under the name of racial wellbeing as a whole, sacrificed themselves to become reproduction machines. Although Ren sided with Zhou's and Sun's Lamarckian emphasis of environmental improvement, he was not particularly convinced by their statements, since neither of them provided any evolutionary theories or statistical studies, whereas Pan's statistics seemed to be more scientifically grounded.[46]

In order to make a thorough evaluation of these debates and a Lamarckian argument, Ren first went back to Darwin's conception of "variation" within species and the principle of adaptation to the environment as the major factor requiring species to change. By demonstrating the interaction of species' adaptation to the environment as the driving force of evolutionary process, Ren reminded us of the importance of environmental factors which deserved their due attention. The statistical studies of eugenic categories such as intelligence, crime rate, and social achievement were of the social categories of human adaptation. Ren diverged from Sun to admit the heredity of instinct and intelligence, but he regarded them to be subject to social conditions for their realization; hence, geniuses and the talented were the products of society in a certain age. Ren pointed out that the statistical studies cited by Pan as often showing a correlation between class and intelligence could in turn attest to evidence of environmental influence. Ren attacked severely Pan's preference for the perpetuation of the upper-middle class and antipathy for social mobility of the lower class. Ren commented that eugenics served the aristocracy in the age of nationalism as well as it served imperialist expansion in the age of international competition. No wonder the eugenics movement coincided with racism, Ren lamented. He also blamed the renowned demographer Chen Changheng (1888–1987) for his hasty acceptance and introduction of eugenics into China without taking into consideration its inherent racist danger.[47]

Ren further questioned the scientificity of eugenics in comparison with other sciences like physics, chemistry, biology, economics, sociology and the like, whose service transcended the boundaries of particular nations and races. He critiqued that, with the feudal features of serving aristocracy, agrarian interests and racial preference, eugenics was not a science but a reactionary imperialist policy. Interestingly, instead of rejecting eugenics, Ren recommended a revisionist environmentalist eugenics, which was not different from euthenics, even under the given fact of the different hereditary personal qualities. Ren suggested on the one hand an improvement of social environment favorable to all classes, and on the other hand research on genetic diseases to clarify their congenital or acquired nature and investigate the way in which society could provide the most congenial surroundings for gifted elements. Ren believed that his reconciliation of the different approaches of Sun, Zhou and Pan was a revisionist eugenic plan devoid of class privilege, racial prejudice and social injustice.[48]

After Zhou, Sun and Ren, other propositions such as dietary improvement, physical education, geographical and meteorological analysis, body hygiene, public hygiene, and economic reform were suggested by scholars and collected in a book entitled *Kexue de minzu fuxing* (Scientific national revival), published by the Science Society of China in 1937.[49] Among these propositions, the most important were those associated with the New Life Movement initiated by Chiang Kai-shek, and the subsequent discursive campaigns of Cultural Construction (*wenhua jianshe*) directed by Chen Lifu from 1934.

The New Life movement consisted of propaganda and legislation aimed at narcotic elimination, body and environmental hygiene, women's education for a war economy, and foreign import reductions, which would purge the unhealthy habits of body and mind in the Chinese people.[50] The focus of the New Life Movement on moral and physical cleanliness linked four Confucian virtues: *li* (propriety), *yi* (justice), *lian* (integrity) and *chi* (shame conscientiousness). Physical reforms concerned gymnastic exercises and the problems of spitting, smoking, prostitution, drinking, opium addiction, and women's indecent exposure in clothing, which defined a bodily discipline and brought forth the mental awareness of a national subject. The definition and embodiment of a national subject as a whole would serve to draw the line between desirable and undesirable elements. Alongside the formation of a national subject and a focal point for national identity, the State was able to exercise power upon each individual body.

However, each individual body would perish after one's lifetime and the result achieved from its moral and bodily disciplines would vanish as well. Hence, Chen Lifu's Cultural Construction campaigns were aimed at the perpetuation of a cultural body from which each desirable individual body could be reproduced accordingly. In his discussion on Cultural Construction, Chen commented that the New Life movement focused on the present, whereas Cultural Construction would look forward to the future. His first priority was the attainment of self-understanding of the Chinese cultural essence—the spiritual, as opposed to Western materiality—from which perspective one could recognize its glorious past and establish national confidence. Chen's second step was to synthesize Western science and material creativity as a new element to complement the old spiritual legacy.[51] These two recipes became the guidelines for cultural-revival discursive practices in the organ journal *Wenhua jianshe* (Cultural Construction), which carried a series of national histories of various aspects of culture, presumably reconstituted with scientific methods.

In January, 1935, the journal featured a manifesto of "Zhongguo benwei de wenhua jianshe xuanyan (Wholesale China-centered Cultural Construction)," signed by ten professors, such as Tao Xisheng and He Bingsong, as a response to Chen Xujing's 1934 declaration of a "Wholesale Westernization" program. This initiated a series of extensive discussions on Chinese culture. Pan Guangdan, even though accepting this manifesto and its implied approach, first contested the meaning of wholesale China-centeredness. He said "*Benwei* (centered on the root) shows the order of the procedures and the priority of emphasis, and indicates a bodily subject (*zhuti*). *Zhongguo benwei*, if translated into English, would be 'Sino-centric' like the old name of China, filled with a connotation of self-conceit." Pan wished the name of *benwei* to signify self-dependence, self-love and self-respect, instead of cultural egotism. He then moved on to comment on these discussions on culture and their proposals to revive Chinese culture. Pan pointed out that the biological factor of heredity was often left aside. For instance, the concept of phenotypic variation was misconstrued as being acquired, and hence reversible through environmental and educational reforms. While the heredity of physiques was recognized, the inheritance of mental and physiological traits was ignored.[52]

Pan suggested examining the Japanese case in order to illuminate the difference before and after the racial/national rejuvenation. Before Japan became a great power, it had been through constant bitter competitions among the archipelago residents such that the cultural institutions of martial spirit and a warrior system, to retain leadership and maneuver, were selected and refined. Without external invasions, Japan's racial stamina was never traumatized. Likewise, before the Manchus ruled China, they had undergone a sequence of positive selection such that they could achieve both military and cultural success in a short period of time. In comparison, China did not experience the same process; consequently, symptomatic characters such as selfishness, inadequate creativity, delicate physique and disjointed organization were selected.

In order to reverse this dysgenic selection, Pan presented a remedial proposition. First, ameliorate people's livelihood, especially by reducing famines and preserving the environment. Second, elevate economic productivity and effect just distribution of wealth, which would guarantee a better average living standard. Third, control the speed of urbanization and the flow of migration from the rural areas to the cities. Urbanization was the source of hygienic problems—such as air pollution, high mortality due to epidemics, sexually transmitted diseases, prostitution, high suicide rates and narcotic abuse—from which city dwellers aged and died prematurely, thus reproducing less. The cities attracted and drew excellent, young elements migrating from their rural homes, and wore them down quickly while impairing the quality of the rural population. In a short time, not only could the urban population not replenish itself, but the rural population would also deteriorate significantly.

Fourth, Pan proposed to remold the family system to foster eugenic ideas and education. The particular ideal family system Pan suggested was one which would not forget society while retaining the individual's selfhood, and would have empathy toward the past while looking forward to the future. Family as a system, in Pan's eyes, defined a person's quality, since it bore one's seed and carried one's blood regardless of one's class and social status. The verse "Do not disgrace what you were born of" from the *Book of Poetry*, the common inscription "Fit for your posterity (*yizisun*)" on household pottery, and the traditional virtues of "benevolence" (*ren*), "filial piety" (*xiao*), "loyalty" (*zhong*), and "forgiveness" (*shu*) were time-honored Chinese eugenic practices passing on through the family system. However, the same eugenic principles had been discovered and studied in the West in recent decades. Pan pointed out that the eugenic emphasis on "*zhong*" (seed) and "*xuemai*" (blood relationship) in the Chinese family could be found in the family genealogies, in which one encountered most often the characters "*gen*" (root), "*yuan*" (origin), "*quan*" (spring), "*mao*" (exuberance) and "*rong*" (glory). These precious eugenic principles embedded in the traditional family system should be restored and preserved in the modern family which would be a combination of best parts from extended family and nuclear family.

Fifth, China should retain the freedom of speech, thought and scholarship, and most importantly the examination system to select talented human resources. The traditional civil-service examination testing of candidates for their intelligence and special talents had the similar merit as I. Q tests for educational purposes. Hence, it could be modified in terms of examination subjects and methods, not necessarily aimed at selecting civil servants. Sixth, one should encourage humanism and racial awareness in education, and diversify its design to enrich each individual's personality. With the definition of Pan's own words of racial awareness (*zhongzu yishi*), Pan's concept of "race" was both cultural and physical consciousness for self-identity. Humanism helped select people from the middle ground rather than from two extremes. Interestingly, Pan gave a sexual example of two presumable extremes, invoking prostitutes as a symbol of indulgence and nuns as one of asceticism, both of which were barren and thus dysgenic. The middle-ground women Pan expected to select in his humanistic education and diversification were those who could be aware of their femininity (*nüxing*), being constituted by what they were in marriage and family, while knowing how to coordinate with and complement their male counterparts. Men's role was to recognize women's different functions and respect their equal lots of contribution to society and nation. The encouragement of racial awareness should not be mistaken as racism, but alertness as to what human beings we would reproduce. Pan saw this biological consciousness as important to add to the family and education reforms.[53] Interestingly, after Pan's proposition, the journal *Wenhua jianshe* (Cultural Construction) submitted in 1936 a prospectus entitled *Yaxiya xueyuan jihua dagang* (the program of Asian Academy). A school of racial improvement (*Renzhong gaizao xueyuan*) was included in the plan to aim at the ethno-racial studies of Japanese, Indians and Chinese in order to prepare a fundamental plan for racial improvement.[54]

Against expectations, Pan's eugenic proposition was different not only from the European versions, but also from the Japanese project I will discuss in the following section. None of the six approaches in Pan's plan could fit into the previous understanding of the eugenic movement to bring forth a legislation of eugenics bills in many parts of the world. In addition, all of his six eugenic prospects emphasizing cultural mechanisms and environmental components yielded to what was called Lamarckism, when the cultural body was perceived as a source of problems for survival and Lamarckism as an antidote to cure this cultural body. In this fashion, Lamarckism offered a solution to the dilemma of how to become fit while remaining "Chinese."

B. JAPAN

Like the Chinese hygienists' inclusive approach to eugenic ideas, the Japanese hygienists appropriated eugenic ideas to justify and further enforce their public-hygiene policies, as shown in the particular case of leprosy studied by Fujino Yutaka.[55] In 1907, the Japanese state sanctioned legislation for leprosy prevention which enforced household registration in order to investigate the total numbers of lepers and thereby, as first priority, to gather together the then homeless lepers into special asylums. Since leprosy was regarded as a chronic epidemic, the final goal of this prevention law was to eventually isolate every leper from Japanese society and to eradicate this national stigma.[56]

From 1920, the hygiene committee sponsored by the Ministry of Home Affairs (*Naimushô*) proposed the establishment of a national leprosy asylum to accommodate the most irreclaimable elements among lepers, who gambled, raped, committed violence, fled and defied management, and to perform sterilization upon the lepers' voluntary request. This proposal was accepted in the 52nd Diet meeting in 1927 and was proceeded on within a schedule of three years. Fujino suggests that this new approach of enforcement had something to do with, in addition to the pressing social concern of the leprosy issue, the American anti-Japanese sentiment and, especially, the impact of the eugenics movement in the 1920s. In 1926, the hygienist Takano Rokurô (1844–), the first chief of the Prevention Bureau in the Ministry of Home Affairs, expressed the eugenic idea of purifying racial blood in leprosy-prevention measures such as the absolute segregation of a voluntary leper colony, a national asylum and the promotion of philanthropic detention-hospitals, which were mostly incorporated into the 1927 proposal submitted to the Diet.

A specialist on leprosy in India, Mitsuda Kensuke (1876–) attending the Third International Convention of Leprosy in 1923, said that it was an inconceivable disgrace that lepers in Japan, a so-called pure-blood nation, outnumbered those of the developed Euro-American countries and even the underdeveloped India, and had the second largest total next to China. In 1928, Mikami Chiyo, once a leprosy missionary, also used the eugenic expression of racial purification to attract public attention to this particular issue and its solution of eradicating leprosy as a Japanese national mission.[57] The above leprosy activists and hygienists all adhered to the eugenic rhetoric of "pure blood" and "racial purification," while acknowledging "leprosy" as a chronic epidemic and nonhereditary disease. The role of politics played in the conflicting mixture of the nonheredity into the genetic concern of hard-line eugenics helps our understanding of "racial hygiene" in a rather literal sense.

In addition to leprosy, the eugenicists often exhorted that sufferers of other malignant nonhereditary diseases, such as tuberculosis, syphilis, mental diseases and feeblemindedness, should be prevented from marriage. With the concern of making a better plan for social control of population quality, the Ministry of Home Affairs considered in 1927, in the process of improving the Leprosy Prevention policy, the option of sterilization for lepers and mental patients as an alternative to marriage restraint. The Ministry received a professional response in favor of excluding lepers as candidates for sterilization, from the Japanese Physicians Association in 1928. However, the then Tanaka Giichi Cabinet member Hatoyama Ichirô (1883–1959) commented on a strong connection between the eugenic concern of compulsory sterilization and the fundamental eradication of leprosy and malicious hereditary criminality. Around the same time the director of the Social Bureau, Okaryû Ichirô, expressed concern about how to reconcile the complaints of birth control being imposed on lepers and mental patients, and the deprivation of their sexual gratification, with the bottom line of not violating the Mendelian Laws. Consequently, in 1929 at the 56th Diet meeting, there was a hearing on an amendment to the Leprosy Prevention Bill. In the hearing, the eugenic approach to leprosy was finalized and the hygienists working for the Ministry of Home Affairs somewhat legitimized sterilization as an established practice.[58]

Moreover, in 1930, there was a rather aggressive proposal that the state should frame a eugenic law to prevent any proliferation and transmission of feeblemindedness, mental diseases, venereal diseases, alcoholism, tuberculosis and leprosy, and mandate a sterilization surgery for these patients and sufferers before their marriage. This eugenic proposal was an early version of the National Eugenic Bill enacted later in 1934, and its approach to legislation conformed to the ways in which the leprosy specialists had been trying to segregate and forestall this disease. This converging effort of leprosy eradication from the eugenicists and hygienists was encouraged. It was considered not ruthless but rather "humanistic" by the world renowned bacteriologist Shiga Kiyoshi (1870–1957), who discovered the dysentery bacterium (later named Shigella) and saw sterilization as a humanistic option for lepers to live a married life without the risk of transmitting their disease to the next generation.[59]

These two converging forces from the eugenicists and hygienists by no means implied that the two groups could be lumped together, as I also point out in the Chinese context. For the hygienists, the most radical and effective medical treatment for any diseases or epidemic was to eradicate the roots, namely, the communicable and transmittable microbes. For the eugenicists, this aggressive medical intervention was merely a passive or negative agenda of the eugenics movement. Hence, this point of convergence is precisely the point of divergence between the eugenicists and hygienists. In the following contrasts and critiques among socialists, sociologists, biologists and eugenicists, we can observe not only nuances among these various Lamarckian groups, but also the shifting position of the eugenicists in their acknowledging the environmental factors along the progression of debates. Among these groups, I will focus mostly on the works of Nagai Hisomu, the chairman of the Japanese Racial Hygiene Society and later of the Japanese Racial Hygiene Association.

Nagai once used August Weismann's work disproving the theory of inheritable acquired characteristics to invalidate that exclusively altering external factors would change heredity. Hence, judged by Nagai, the conventional measures of medicine to cure diseases and improve human health, and education, religion and law to rectify one's behavior and revive one's morality, were useless to the cause of human betterment.[60] In 1916, he introduced eugenic technology by using the evident effect of horticulture and goldfish breeding to make a convincing case for possible human improvement from man-made selection. He gave an example of Max von Pettenkofer, who swallowed cholera germs in order to discredit his opponent Robert Koch and Koch's discovery and theory of a microbial cause for cholera. According to von Pettenkofer, the cause of disease came from a body susceptible to bacteria, rather than from the bacteria themselves; therefore, in order to prove his own disease hypothesis he swallowed the cholera germs and suffered merely minor diarrhea. Nagai used von Pettenkofer's example to argue that certain people were more susceptible to tuberculosis than others, illustrating the innate differences among everyone's physical constitution. Although internal and external factors were both important to our understanding of disease, weighing both factors in terms of magnitude one found the internal one far more essential than the other one. Thus a Lamarkian emphasis on environmental improvement could never appropriately improve racial stock.[61]

Nagai's unreserved rejection of the Lamarkian emphasis on external improvement reflected the development of genetics at the time when Mendel's laws were recovered and accepted in 1900–1915. In addition to warning against consanguineous marriage, Nagai also considered nonsensical and scientifically baseless the conventional belief in prenatal care and education (*taikyô* or in his own words, maternal impression) and telegony, that the maternal body would pass on her experiences, consciousness and impressions to the fetus during the pregnancy.

Later, in his works of the 1930s, when the Lamarckian idea of inheritance of acquired characters was appropriated respectively by Paul Kammerer in 1926 and Trofim Lysenko from 1932 on (to 1965), Nagai, while still maintaining his previous Mendelian position, started to address seriously and engage actively in the discussion of environmental factors. Within the scientific network, Nagai's seemingly hard-line rhetoric of Mendelism differed essentially from the views of the Lamarkian biologist Oka Asajirô (1866–1944) and a group of young Marxist biologists who established the Society for the Study of Materialism (*Yuibutsuron kenkyûkai*) and established the journal *Yuibutsuron kenkyû* (Studies in materialism) in 1932. These young Marxist biologists included Ishii Tomoyuki, Ishihara Tatsurô, Hashikake Akihide, and Ishii Tomoyuki, and their arguments were merged and synthesized into a monograph *Seibutsugaku* (Biology) edited by Ishii and Ishihara and published in 1935.[62]

They repudiated the conventional simplistic and mechanistic approach of the biologists to the interpretation of social problems. They demarcated human beings from other general living creatures by three criteria, genetic characters including physiological and morphological ones, environment, and adaptation. For living creatures in general, since genetic characters were not entirely passed on from the parent generation to the offspring, the natural environment played a key role in the process of evolution, in which the links among living creatures were rather weak, i.e. their adaptation to the environment was rather self-concerned, passive and direct. As for human beings, in addition to the parents' genetic characters, men inherited other material and spiritual legacies, and along with the natural environment the social environment was very complicated. In the process of interaction with Nature, men evolved variation and accordingly dominated Nature by following their own practical purposes. They commented on the hasty biological application of eugenics to human society without clarifying scientifically the real biological cause of malignant elements endangering social existence, and more importantly the formation of social relations leading to reproduction as such. They posed a deep suspicion of the activities of racial hygiene in the recent years.[63] For them, the dialectical interaction between environment and individual organism determined the function and morphological variation of organisms. However, these complex evolutionary processes were simplified and reduced by Mendelism to a specific combination of genes and their changes within a cell.[64]

Despite Oka's and the young Marxist biologists' support of the tenet of inheritance of acquired characters, most geneticists accepted Mendelism, especially after W. L. Johannsen's theory of pure lines was widely accepted in the 1910s.[65] The young Marxist biologists did not sustain their Lamarkian impact and influence in the intellectual and social domains, because the scientific socialist movement in Japan had been thoroughly crushed about 1930. The prominent Marxist biologist Yamamoto Senji (1889–1929), who confronted the eugenicists on the issue of the birth-control campaign, was assassinated in 1929, other young Marxist biologists' uncritical acceptance of Soviet theories led to a tendency toward dogmatism,[66] and the Society for the Study of Materialism was disbanded in February, 1938. In addition, the introduction of Lysenko's criticism of orthodox genetics by Yasugi Ryûichi (1911–) masqueraded as a form of merely reviewing Soviet literature. Lysenko's theory, as Yasugi presented it at the time (1937–42) or even later, did not produce sufficient evidence to negate or reconcile Mendelian experimental genetics.[67]

Other than the diametric oppositional confrontation between Marxist biologists and Mendelian geneticists, some socialist-oriented eugenicists tried to reconcile these two different approaches into a comprehensive eugenic program. After Gotô Ryûkiji established the journal *Yûseigaku* (Eugenics monthly) and the Nihon yûseigaku kyôkai (Japanese Eugenics Association) in 1924, he and another eugenics activist, Ikeda Shigenori, helped set up the Nihon yûsei undô kyôkai (Japanese Eugenics Movement Association) and the journal *Yûsei undô* (Eugenics Movement) in Tokyo in 1926. Ikeda's approach was a rather extensive social engagement of eugenic ideas and was aimed at a large audience of different social strata in order to promote the eugenics movement into a national, popular and patriotic (*kokumin, minshû* and *aikoku*) movement. The *Yûsei undô* campaign launched by Ikeda and Gotô went beyond the hard-line genetic stance and inclusively encompassed: 1) marriage selection based on genetic principles, in order to curtail inferior progeny and increase better offspring, through the reformation of conventional marriage customs and law; and 2) educational improvement of social heredity by the euthenic measures of medical care, social medicine (health care, hygiene and preventive medicine), and mind-body discipline (gymnastics, sports and spiritual cultivation).[68]

This encompassed vision of eugenics was also cherished by the prominent sociologist Takebe Tongo (1871–1945), who tried to integrate the genetic factors of the individual organism into a social body. In his 1931 work *Yûseigaku to shakai seikatsu* (Eugenics and social life), he extended the meaning of eugenics as a prenatal genetic improvement of the individual organism, into an enveloped application on a higher evolutionary form. This form was the social body, in which a eugenicized individual located his or her meaning and value of life—namely, self-sacrifice for family and nation—and harmonized his or her extended social relations as a whole. Takebe perceived eugenics as an evolutionary drive to help evolve a lower form of individual organism into a higher form of personified social organism. Hence, his approach to the eugenics program had no longer merely a physiological, but also a cultural and spiritual consideration—a eugenic action to unify mind and body.[69]

While Takebe was aggrandizing eugenic signification into the social space such as home and family, school and the public (*seken*), and in the social milieu of reprehensible customs of prostitution, alcoholism, smoking, abortion, infanticide and contraception, Ikeda was trying to locate a social foundation in the local polity to embody his eugenic ideas. In his article "Yûsei undô to chihô jichitai" (The eugenics movement and a local self-government system), he argued that the "better society" to which the eugenicists looked forward was the realization and popularization of a well-regulated constitutional government, without which the eugenics movement could not operate on a local basis. Moreover, a local self-government system was indispensable to engage extensively, directly and bindingly with social reality in the eugenics movement.

Man can not survive outside of the social life, Ikeda continued. It is normal for human beings to live in a sometimes large and sometimes small collective life, of which the former would be the nation-state (*kokka*) which unifies various groups, and the latter the local self-government. The local self-government, from the perspective of the nation-state, was merely as a cell to a body. Similar to the way of building a house step by step, the eugenics movement aimed at safeguarding the health of such cells, and finally the health of the social body and the national body.[70] A healthy local self-government, as Ikeda envisioned it, would not only institute physical discipline, order, training and control, but would also exalt independent spirit and regional autonomy. As the self-government functioned properly, the future of each locality was pictured by Ikeda as a worldly paradise of well-connected transportation, adequate hygienic facilities, fully-equipped sewerage and irrigation systems, remarkable educational facilities, increased productivity, and most importantly an abundant, happy, cheerful and healthy life.[71]

According to Ikeda, there were several viable conditions for Japan to develop the eugenics movement. Geographically, Japan as an archipelago nation is a country contained by its natural boundary and inhabited by a homogeneous "Japanese race." Therefore, such a nation was an ideal location to implement a constitutional state, and the system of self-government was a perfect form of collective life. Historically, the kinship system in Japan was well developed and connected on the village level. From the perspective of the eugenic process of purifying the Japanese racial blood, the self-government of each village provided a comparable network for marriage selection. This overlapping networking would not only bind cohesively the people with the emperor and the Imperial family, but also strengthen markedly the ties among citizens. In no time, Japanese racial quality would be sophisticatedly refined. Compared with the difficulties in the metropolis for the eugenics movement, Ikeda envisioned a brighter future in the local self-government system.[72]

Ikeda's blueprint for this collective project was endorsed by a renowned politician, Tsurumi Yûsuke (1885–1973), the son-in-law of Gotô Shimpei (1857–1929), the director of the Public Health Bureau of the Home Ministry, the chief Civil Administrator of the Government-General in Taiwan and then the president of the South Manchuria Railway Company. Tsurumi, a Diet member and then minister of Health and Welfare in the Hatoyama Ichirô Cabinet in 1954, commented on Ikeda's campaign as an ideal collective movement that focused on both individual physical quality and communal spiritual value. Tsurumi also mentioned his own idea of designing a similar project, which he called "Shu no undô (the movement of species)" with a slightly different content.[73]

Ikeda's idea of unifying various self-governments as cells into a homogeneous aggregate body within the form of a nation-state that could harmonize and coordinate with each body-part's movement, not only won the support of politicians like Hatoyama and Tsurumi, but also broadened the hardline interpretation of the eugenics movement. It provided an expansive and encompassing ideological rhetoric to define the individual heredity and body as part of the nation-state body. This stance to involve the state's representation, intervention and mediation integrated the agendas of both the hardline eugenicists and the hygienists when the Sterilization Bill underwent a process of national legislation.

When the Nihon Minzoku Eisei Gakkai (Japanese Racial Hygiene Society) re-organized into the Nihon Minzoku Eisei Kyokai (Japanese Racial Hygiene Association) in 1935, one of the Association's emergent goals was to urge the enactment of the Sterilization Bill. The Bill was first drafted in 1934 by the Society and was accepted by Diet member Yagi Itsuro and by the director of Hygiene Bureau in the Home Ministry. Yet it did not then make any further progress. In 1935, the Bill was accepted and presented directly by another Diet member, Arakawa Gorô, to the 67th Diet meeting, but was shelved. The bill Arakawa presented used the rhetoric of pure versus poisoned blood to make a strong case for imposing marriage restriction and voluntary sterilization upon the sufferers of venereal diseases, tuberculosis and leprosy, in addition to other genetic diseases like feeblemindedness, epilepsy, schizophrenia, alcoholism, mental illness, hysteria, pathological crimes, and hereditary blindness and deafness.[74]

In 1937, the eugenicist Nagai Hisomu chaired the Eugenics committee within the National Physical Strength Investigation Commission (*Kokumin tairyoku mondai chôsa iinkai*) of the Japanese Learning and Study Advancement Society (*Nihon gakujutsu shinkôkai*) and submitted to the cabinet a proposal to establish a eugenics bureau within the bureaucracy system. The Cabinet granted the proposal and instituted a eugenics division in the Ministry of Social Health (*Shakai hokenshô*, later changed to *Kôseishô*). This eugenics division conformed to Ikeda's approach to locate the institutional bases of eugenics within the state's organs, and it became a forerunner of the later Ministry of Health and Welfare (*Kôseishô*). It was not until 1938, when the Kôseishô was established, that the Kôseishô-sponsored-and-named National Eugenic Bill (*Kokumin yûsei hôan*) was introduced before the 75th Diet and passed on March 8, 1940.[75]

The National Eugenic Bill was targeted exclusively at the category of genetic diseases and deformities requiring sterilization, and it left out leprosy because of its nonheredity. But it recognized the necessary medical treatment of sterilization based on a doctors' diagnosis, and the consensus of the Kôseishô officials together with the central and the local eugenic investigation committees. In this case, doctors still could perform sterilization under the category of medical treatment upon patients of venereal diseases, tuberculosis, leprosy and other chronic diseases. In addition, the central and the local eugenic investigation committees preserved their authority to issue any order of compulsive sterilization whenever circumstances were appropriate.[76]

In this process of national legislation to produce a National Eugenic Bill, we observe a turnaround of both eugenicists and hygienists, who found again a meeting ground. In 1933, Nagai Hisomu wrote a long treatise on physical training and education (*Taiiku ron*) to promote the positive value of physical training and exercise as an appropriate way to improve the health of the human body and mind. Together with detailed instructions on how to exercise, and a discussion of how heart and lungs function in blood circulation and breathing, Nagai did not forget to remind his readers of the importance of physical education in contributing to Japanese physical and mental strength for the sake of their colonial enterprise.[77] His writing on food and nutrition expressed his concern as to insufficient intake of animal protein and calories in the Japanese diet as opposed to that of European countries. Based on the dietary studies of the hygienist Toda Shôzô (1885–1961, a Racial Hygiene Association member), Nagai suggested an increase of milk, poultry, meat and especially seafood to supplement the conventional Japanese principal diet of rice and vegetables. Instead of the neo-Malthusian and Margaret Sanger's proposition of a birth-control campaign to thwart the threat of population increase and scanty food supply, Nagai suggested the source of staple food supply to be located in the Japanese colonies in Korea and Taiwan as a positive solution.[78] After the discussion of food and nutrition, Nagai continued to rationalize daily hygiene (Nijûyo jikan no eiseiteki gôrika), in which he focused on a disciplined life-routine free from toxicants of tobacco, opium and alcohol in order to present a sensible and healthy daily life.[79]

Moreover, in the article "Minzoku eisei to shakai eisei no akushu" (A handshake of racial hygiene and social hygiene) published in 1936, Nagai recognized the need of euthenic improvement to remove environmental poisons to lower the infant mortality rate. Seeing that male infant mortality was generally higher than female infant mortality, Nagai reconciled that a viable environment could not only prevent any lethal toxins and reduce infantile mortality, but could also help cultivate the well-born infants' phenotypic expression. Therefore, a child-care program would not obstruct but encourage the selection process. Nagai admitted that the earlier conclusion of dichotomies between social hygiene and racial hygiene, and between euthenics and eugenics, developed by him and his eugenicist colleagues, should now be revised and that the hygienists and eugenicists should shake hands in the face of the important childcare problem.[80]

This diametric shift in Nagai's position conveyed the merger of the trajectories between the Japanese eugenics movement and the social hygiene campaign, which in turn unfolded new pages in the history of hygiene. The hygienist Teruoka Gitô (1889–1966, a medical doctor, a student of Nagai Hisomu at Tokyo Imperial University and a member of the Racial Hygiene Association, with a specialty on labor science) provided a narrative of historical development from individual hygiene in the 18th century into the present stage of racial hygiene.[81] The rise of the Industrial Revolution, in this narrative, served as a locomotive driving individual hygiene into a larger form of social knowledge. It not only contained a nucleus of principles based on physics and chemistry, but also was wrapped with branches of biology such as morphology, anatomy, pathology, etiology, bacteriology and histology. (As I discussed in chapter two, hygiene was taken to be a composite whole of these various branches.) The recent progress of genetics and its direction toward social control in the so-called racial hygiene was leading hygiene to a new way of thinking about its connection to social policy. According to Teruoka, hygiene should belong to the social sciences. Teruoka repudiated the conventional understanding of hygiene among numerous hygienists, as one of the hard-core sciences based on physics and chemistry, without recognizing its social application and its integration with other social-science methodologies. Teruoka said that hygiene was not a subject with definitive and fixed scope and method. This is because it evolved and changed over time whenever it received impact from any up-to-date development of knowledge, first from cytology and pathology, then bacteriology, and finally the recent genetic and evolutionary theories on germ plasm and environmental conditions.

Different from the earlier isolated stance of hygiene, some hygienists including Teruoka were starting to make efforts to integrate medicine with other human sciences. Teruoka further elaborated this new approach through a thematic differentiation in the scientific changes of three historical phases, that is, personal/individual hygiene, social hygiene and racial hygiene. According to Teruoka, personal/individual hygiene studied the object of individual health and the influences received from environmental impact, and social hygiene prioritized the majority of individuals aggregated by the moral and social principles in order to maintain the metabolism and replenishment of the social community (*shakai kyôdôtai*). By virtue of class differences and vested interests, there was by no means a sweeping concord between these two different approaches; therefore, social hygiene should enlist the resources of human and social sciences to help smooth out the difficult problem of conflict and incompatibility. Social hygiene, as defined by Teruoka, should integrate experimental methods derived from the natural sciences with statistical procedures in the social sciences to probe into various mechanisms of economy, morality and spirituality, in addition to other environmental conditions, that contributed to and determined the wellbeing of the community.

As for racial hygiene, this newest branch of hygiene targeted race as a base unit for sexual continuation (*sei no renzoku*) and its goal for racial betterment was to develop and popularize the future ethno-hygienic culture. Like personal/individual hygiene and social hygiene, racial hygiene recognized the function of environmental forces that would shape and select individuals and their progeny with certain desirable qualities in the process of human reproduction. Although Teruoka expressed the difficulty of the genetic application of racial betterment when it encountered the barriers of social and cultural habits, he nevertheless believed in the development of racial hygiene as a right track for the optimum wellbeing of the social and racial body in a physical sense.

The dynamics of Teruoka's narrative was the transforming power generated from the evolutionary narrative not only to sublimate the interest of individual health and life perpetuation into a deep form of national concern on the racial body, but also to provide the epistemological "telos" of totality or closure for the discipline of hygiene. To a certain extent, Teruoka's delineation of racial hygiene expanded the scope of Nagai's eugenic approach, helped the eugenicists to claim the disciplinary niche of conventional hygiene, and anticipated the later embodiment of the *Kôsei* (public health and social welfare) enterprise from the deeply-implanted principle of totality (*zentai shugi*).

Placing Teruoka's evolutionary narrative against the historical context of Japanese physicians' attempts to combat epidemics such as smallpox and cholera, we find an innocent history of physicians' search for an effective agency to quarantine transmittable diseases, and their medical ethics of compassion and objective treatment of suffering patients, regardless of their social class and wealth. The different phases of this search were from *yôjô* (transliterally, nourishing one's life, or preservation of one's health), to *eisei* (protecting life, or hygiene and sanitation), *yûsei* (ameliorating life, or eugenics and racial hygiene) and *kôsei* (enriching life, or public health and social welfare).

Yôjô, as expounded in the traditional manuals such as Kaibara Ekiken's *Yôjôkun* (Rules for nourishing life, published around 1713) and Uratani Yoshiharu's *Yôjô no susume* (An exhortation toward nourishing life, 1876), focused on the subjects of diet, exercise, air and breathing, sexual intercourse, control of desire, bathing, clothing, food preparation, mental balance, pregnancy and childcare. The knowledge of *yôjô* was aimed at spiritual and physical well-being of one's own body, and hence emphasized self-cultivation, self-mastering and self-hygiene of the individual body. Accordingly, the medical treatment of illness centered on the self-contained system of the individual body and its interaction with the natural world. Even though the Western knowledge of anatomy was incorporated into *yôjô* manuals around this time, they did not substantially reflect the epistemological shift of the subject locus from the individual body and household to a greater social body, until the new method of extensive sanitation was introduced.[82]

In combating devastating epidemics like cholera, dysentery and typhoid fever seething in the last decades of the 19th century, *yôjô* knowledge and care of the individual body became insufficient, and gave way to the much more comprehensiveapproach of *eisei*. *Eisei* was an elegant word first used in this way by Nagayo Sensai (1838–1902, the first director of the Hygiene Bureau), from the chapter *Gengsangchu* in the Chinese classic *Zhuangzi*. It meant health protection, and, as used for the name of the Hygiene Bureau since 1875, extended the Bureau's responsibility from medicine to nationwide sanitation.83 The approach of *eisei* sought to bring community hygiene to public awareness as every citizen's duty. *Eisei*, as Nagayo's successor Gotô Shimpei later interpreted it in his work "Kokka eisei genri (the principles of national hygiene)" became the *eisei* of the nation-state (*kokka*). He saw the nation-state as an analogously sublimated form of the human body. Upon such a form the evolutionary force could manifest itself in the change from ignorance to enlightenment, from shamanistic superstition to sanitation, from treatment to prevention, from prolonging life to enriching life, and from the savagery to civilization. In order to function and achieve perfect physiological harmony, the nation-state as an organism would coordinate various organs in a time of peace, while in a time of turmoil it would defend its life (the majority's *eisei*) and forsake individual ones.84 It was this very principle of totality—of forsaking individual life and subsuming it into a meaningful whole—that brought forth the historical synthesis of individual body and social body into the national/racial body embracing the approaches both of public hygiene and of racial hygiene.

CONCLUDING REMARKS

The role of Lamarckism as a cultural logic playing in the Japanese context helped provide an analogous extension from the individual body to a greater national/racial body, in that undesirable elements were justifiably removable for the sake of the wellbeing and health of *zentai* (totality or literally the body as a whole). The irony implied herein was the removal of leftist socialism, like the eradication of leprosy, by the rightist socialists on the basis of the same Lamarckian emphasis. Hence, the mind in the *zentai* body, manifested as the state operating in the nation, became a homogenizing unity to coordinate harmoniously on everybody's part. In comparison, the Mendelian approach in Chinese eugenics was not only vilified by socialists of both the left and right, but also was transformed and succumbed to the concern of generational transmission in the Chinese cultural body. The shift of attention directed Chinese eugenics down the route of cultural renaissance and rejuvenation, rather than the route to eugenics bills legislation.

In the Japanese eugenics journal *Yûseigaku*, a 1936 Japanese travel note from North China, before the outbreak of the second Sino-Japanese war, reported that the Chinese eugenics movement was encountering a hopeless impasse, because the Chinese society was sick in the midst of fierce turmoil. There was no strong and critical middle class in Chinese society to defend the benefit of the majority. The elite followed mechanically what they learned from their overseas education, mainly from Britain and America, without relating the knowledge to the reality of people's benefit. The report went on to say that without integration and order, the society was not able to define or design the desirable type of human being in order to examine and therefore differentiate the good from the inferior elements in every individual's physical and mental qualities.[85] This report flashed historical insight on the outcomes of debates in both the Japanese and the Chinese eugenics movements: whereas the Japanese had achieved and exhibited efficiency in the unity of mind and body, the Chinese were debating how to put together a healthy cultural body, with the function of its mind still in question.

The historical excavation of these two processes of scientific debate not only sheds light on the social forces in action in Japan and China, but also alerts us to the politics of scientific interpretation which is often involved in the struggle over meaning and values. The politics of eugenics was not a mere conservative or rightist phenomenon. What had been said in the various debates should not be stigmatized as a whole by its historical outcome, and be buried in oblivion. The debates are instead the discursive resources to tackle present and future problems.

NOTES

[1] This ideological association of Lamarck's name was stigmatized by the extreme cases of the Kammerer and Lysenko affairs that had less to do with Lamarck himself than with the political appropriation of Lamarck's ideas. In 1926, the Austrian biologist Paul Kammerer, or his assistant, injected India ink under the "midwife" toad's skin and presented it as evidence of the inheritance of acquired characters. It was soon discovered to be a forgery and thereafter Kammerer shot himself. For Lysenko's political endorsement of the slogan of inheritance of acquired characters and the Russian eugenics movement, see Mark Adams' "Eugenics in Russia, 1900–1940," in *The Wellborn Science* (Oxford: Oxford University Press, 1990), pp. 153–216.

[2] Nancy L. Stepan, *"The Hour of Eugenics" Race, Gender, and Nation in Latin America* (Ithaca: Cornell University Press, 1991), p. 81

[3] "Eugenics in Brazil: 1917–1940," in *The Wellborn Science*, p. 121.

[4] Harry D. Harootunian, "The Functions of China in Tokugawa Thought" in *The Chinese and the Japanese: Essays in Political and Cultural Interactions*, ed. by Akira Iriye (Princeton: Princeton University Press, 1980), pp. 9–36.

[5] See Stefan Tanaka, *Japan's Orient: Rendering Pasts into History* (Berkeley: University of California, 1993).

[6] Tetsuo Najita and Harry Harootunian, "Japanese Revolt Against the West: Political and Cultural Criticism in the Twentieth Century" in *The Cambridge History of Japan*, Vol. 6, (Cambridge: Cambridge University Press, 1988), p. 712.

[7] Tanaka's article "Ningen honshitsu no kaisen ga kyumu (A pressing need for the improvement of human qualities)" was quoted in Suzuki Zenji, *Nihon no yûseigaku: so no sishô to undô no kiseki* (Japanese eugenics: its ideas and locus of movement) (Tokyo: Sankyô, 1983), p. 101.

[8] Ibid., p. 102.

[9] Ibid., p. 115.

[10] Carol Gluck, *Japan's Modern Myths: Ideology in the Late Meiji Period* (Princeton: Princeton University Press, 1985), pp. 120–121.

[11] Suzuki Zenji, *Nihon no yuseigaku*, p. 116.

[12] This metaphor was also used by another eugenicist, Gotô Ryûkiji, who established the journal *Yûseigaku* (Eugenics monthly) in 1924, in his article "Yûseigaku to wa don na gakumon ka" (What kind of knowledge is eugenics?) in *Yûseigaku*, vol. 3, no. 3 (1926), pp. 17–19.

[13] Nagai Hisomu, *Yûseigaku gairon* (Tokyo:Yusanko , 1936), pp. 4–5.

[14] Ibid., p. 6.

[15] See Lung-Kee Sun, "The Presence of the Fin-de-Siècle in the May Fourth Era" in *Remapping China: Fissures in Historical Terrain* (Stanford: Stanford University, 1996), pp. 194–209.

[16] According to Rebecca E. Karl's posting on the listserve H-Asia in September, 1997, in the 15 Dec. 1896 issue of the *Shiwu bao*, an essay from the *Zilin xibao* [North China Daily News] was reprinted; it proclaimed that there were "four sick people of the world"—Turkey, Persia, China, Morocco (Tianxia sibingren). Shortly after, these were whittled down to two and renamed. This was in a preface written by Liang Qichao to a book on the Russo-Turkish War of 1877, which was also published independently as an essay in the 51st issue of the *Shiwu bao* (entitled: "E Tu zhangji xu"). Liang states that "Westerners often say that there are two sick men of the East: China and Turkey." He goes on to describe, in the context of exhorting people to care about foreign affairs, what this means for Turkey and for China. He says that Turkey, "previously a big and powerful country" spanning 3 continents, today "has become like a strand of hair that has nearly snapped." What prevents it from snapping, Liang says, is European pre-occupation with the second sick man of the East, China. Also in the beginning of 1898 in his book eventually submitted as a memorial to the throne during the 100–days reform, Kang Youwei wrote that Europe regards China as the "Turkey of the East." (In his Tujue xiaoruo ji [Demise and Weakening of Turkey]: "tujue" here is used to refer to the Ottoman Turks.) Other Western newspapers in Shanghai also frequently used the locution in reference to the Chinese—the missionary journal? "Wanguo gongbao" is another prime source of this locution. With time, Turkey was sometimes dropped from the locution, and China stood alone as 'sick man of the East'. This may also have been reinforced from/in Japan, but its genealogy seems more clear via the Western/ Turkey connection.

[17] See the prefaces by Li Jinghan and Pan Guangdan himself to his own book, *Minzu texing yu minzu weisheng*, originally published by the Commercial Press, Shanghai, 1937, now collected in the Pan's Collection, (Beijing: Beida, 1995), volume 3, pp. 5–26. The term "mirror" was used by Li Jinghan, a renowned sociologist, in his preface.

[18] Ibid., pp. 178–203.

[19] Lung-Kee Sun, "The Presence of the Fin-de-Siècle in the May Fourth Era," p. 202.

[20] Ibid.

[21] See Laurence A. Schneider, *Ku Chieh-kang and China's New History: Nationalism and the Quest for Alternative Traditions* (Berkeley: University of California Press, 1971), pp. 263–269. Gu, from the late 1920s, emphasized the potential physical

contributions of non-Han people, and became interested in their future role in Chinese civilization. Pan's position of inter-ethnic marriage and his shift to ethnological studies of minorities was similar to Gu's.

[22] Shimizu Katsuyoshi, *Nihon kôshû eiseishi* (The History of public health, the early half of the Shôwa period), (Tokyo: Fuji shuban, 1989), p. 68. For a detailed and thorough account, see William Johnston's *The Modern Epidemic: A History of Tuberculosis in Japan* (Cambridge: Harvard University Press, 1995), chapters 2 and 3.

[23] Shimizu, p. 118, and Herbert D. Lamson's *Social Pathology in China: A Source Book for the Study of Problems of Livelihood, Health and the Family* (Shanghai: The Commercial Press, 1935), p.301. Lamson pointed out that the actual number of lepers in Japan was much higher, about forty to sixty thousand then. This book was written by Lamson as a textbook of sociology while he was teaching in the University of Shanghai, and was translated into Japanese by Maejima Yutaka in 1941.

[24] Shimizu, p. 131 and pp. 148–9.

[25] Lamson, *Social Pathology in China*, p. 328.

[26] Ibid., p. 355. The rates were not based on total armies, but upon the numbers coming to hospitals for disease or wounds. Lamson used these figures to show the need for medical prophylactic treatment, for the greater control of prostitution, and for improved discipline in the armies of China.

[27] Ibid., p. 410.

[28] In *Zhonghua yixue zazhi* (the *National Medical Journal of China*), vol. 2, no. 2 (1916), pp. 49–58.

[29] Both Wu's "Duiyu zhongyang yiyuan zhi ganyan (Comments on the establishment of the Central Hospital in Beijing)" and Yu's "Hunyin yu jiating zhi guanxi (On the relationship of marriage and family hygiene) are juxtaposed side-by-side in the *National Medical Journal of China*, see vol. 3, no. 4 (1917), pp. 1–6.

[30] Ka-che Yip, *Health and National Reconstruction in Nationalist China: The Development of Modern Health Services, 1928–1937* (Ann Arbor: AAS monograph and occasional paper series, no. 50), p. 47.

[31] Even though Hu did not indicate the source of original work, from the footnote in Pan Guangdan's article "Yousheng gailun," we know that Hu was translating Charles B. Davenport's *Heredity in Relation to Eugenics* (1911). See Pan's *Collection*, vol. 1, p. 263.

[32] Hu Xuanming, "Hunyin zhesixue" in *Zhonghua yixue zazhi* (the *National Medical Journal of China*), vol. 5, nos. 3 and 4 (1919), pp. 144–148, and pp. 216–224; vol. 6, nos. 1, 2, 3 , 4 (1920), pp. 50–57, pp. 121–132, pp. 174–187, pp. 256–268; vol. 7, no. 1 (1921), pp. 60–71.

[33] Hu Dingan's own memoir article "Guchui yixue geming de huiyi" (Memoir on the campaign for the medical revolution) in *Zhuanji wenxue*, vol. 13, no. 5 (1968), p. 38.

[34] Hu Dingan, "Minzu de liliang zai guomin jiankang (The strength of a nation relies upon the health of the people)" in his collection of medical commentaries *Hu Dingan yishi yanlunji* (Shanghi: Shangwu, 1935), pp. 63–70.

[35] See Shirley S. Garrett, *Social Reformers in Urban China: The Chinese Y.M.C.A., 1895–1926* (Cambridge: Harvard University Press, 1970), pp. 140–148.

[36] Kimberly A. Risedorph, "Reformers, Athletes, and Students: the YMCA in China, 1895–1935," (Dissertation, Washington University, 1994).
[37] Pan's propagation of eugenics ideas and his affiliation with the YMCA have not appeared in any studies on the YMCA so far, and when I interviewed with Pan's daughter, Pan Naimu, she said Pan was a Christian initially and later he changed his mind. It seemed that Pan Naimu did not want to pursue this topic and I speculate that it has something to do with Pan being accused of being a counterrevolutionary and rightist, and being mistreated to death during the Cultural Revolution.
[38] On this particular fundamental issue Pan Guangdan stood on the side with the geneticists; Pan also had a follower, Xu Zhenhua ,who wrote a long article, "Youshengxue yu zhongguo" (Eugenics and China) in *Tung-chi Acta Medica* (Tongji yixue quarterly), vol. 1 (1931), nos. 1 and 2, pp. 16–93.
[39] Both Pan's and Zhou Jianren's articles were published in *Dongfang aazhi* of 1924 and 1925, and now both of them and Pan's response to Zhou's critique are all collected in the first volume of Pan's *Collection.* Pan's first article, vol. 1, pp. 266–288.
[40] Galton, *Hereditary Genius* (London: the Fontana Library, 1962), p. 390.
[41] Zhou's article, vol. 1, pp. 306–315.
[42] In the political spectrum, Pan was an independent non-partisan. It was said that Pan was the next target after the Nationalist Party assassinated Wen Yiduo and Li Gongpu because of their sympathy toward the Communist Party.
[43] "Tantan youshengxue" in *Qingnianjie,* vol. 1 (1931), no. 3, pp. 1–11.
[44] Sun's article "Wenhua yu youshengxue" (Culture and eugenics), and Pan's response "Yousheng yu wenhua" (Eugenics and culture), and Sun's response to Pan's response "Zailun wenhua yu youshengxue" were all published in *Shehui xuekan,* vol. 1 (1929), no. 2.
[45] Ye was a materialist first and foremost. For more biographical details of Ye, see Joshua A. Fogel, *Ai Ssu-ch'i's Contribution to the Development of Chinese Marxism,* (Cambridge: Harvard University Press, 1987), pp. 27–30.
[46] See Ren's article, "Ping Youshengxue yu huanjinglun de lunzheng" in *Ershi shiji,* vol. 1, (1931), no. 1, pp. 57–124.
[47] Ibid.
[48] Ibid.
[49] In addition, there was Zhang Junjun's *Minzu suzhi zhi gaizao,* focused on Chinese dietary improvement and physical education, (Chongqing: the Commercial Press, 1944).
[50] Arif Dirlik, "The Ideological Foundations of the New Life Movement: A Study in Counterrevolution" in *Journal of Asian Studies,* vol. 34 (1975), no. 4, pp. 945–980.
[51] Chen Lifu, "Zhongguo wenhua jianshe lun" in *Wenhua jianshe,* vol. 1 (1934), no. 1, pp. 11–16.
[52] Pan Guangdan, *Minzu texing yu minzuweisheng,* in Pan's *Collection,* vol. 3, 31–39.
[53] Ibid.
[54] This prospectus was drafted by Fu Youren in *Wenhua jianshe,* vol. 2 (1936), no. 5, pp. 26–28.

[55] Fujino Yutaka, *Nihon fashizumu to iryô: hansenbyô o meguru jisshôteki kenkyû* (Japanese fascism and medical care: an evidential investigation in connection with leprosy) (Tokyo: Iwanami shoten, 1993).
[56] Ibid., pp. 8–14. Before the enactment of this legislation, lepers were the most unwanted social elements and were either neglected by their families or sent to missionary asylums. In the process of enactment and hearings in the Diet, most legislators regarded foreign missionary aid to the Japanese lepers as an unbearable national stigma.
[57] Ibid., pp. 29–33.
[58] Ibid., pp. 35–35.
[59] Ibid., pp. 38–38.
[60] Nagai Hisomu, *Seimeiron*, (Tokyo: Rakuyôdô, 1915), pp. 340–351. I am using the tenth edition; the first edition was published in 1913.
[61] Nagai Hisomu, *Seibutsugaku to tetsugaku to no sakai*, (Tokyo: Bunka seikatsu kenkyûkai, 1926), pp. 276–283. The one I am holding is the 1926 tenth edition; it was first published in 1916.
[62] See Nakamura Teiri, "Marixism and Biology in Japan" in *Science and Society in Modern Japan: Selected Historical Sources*, edited by Nakayama Shigeru, David L. Swain, and Yagi Eri, (Cambridge: The MIT Press, 1974), pp. 253–269.
[63] Ishii Tomoyuki and Ishihara Tasuro, *Seibutsugaku* (Tokyo: Sanryû shobô, 1935), pp. 96–118. Interestingly, Ishii and Ishihara also criticized Oka Asajiro for his biological interpretation of society.
[64] Ibid., pp. 269–275.
[65] Suzuki Zenji, *Nihon no yûseigaku*, p. 79.
[66] A renowned evolutionary biologist, Koizumi Makoto, critiqued in 1932 that the *Yuiken* study group cited and quoted uncritically Marxist literature and arguments in their writing, without completely digesting and demonstrating their own ideas. See Nakamura Teiri, "Marixism and Biology in Japan," p. 257.
[67] Ibid., pp. 255–261.
[68] Suzuki Zenji, *Nihon no yuseigaku*, pp. 116–117.
[69] This is a rough abstract of Takebe's approach in his introduction to his book. His idea of action unifying body and mind was derived from the influence of Confucian Chinese scholar Wang Yangming. For the connection of his sociological thoughts to Comtian positivism and Spenserian evolution of the social organism, see Saitô Shôji, *Ninhon shakaigaku seiritsu shi no kenkyû* (Tokyo: Fukumura, 1976), pp. 219–223.
[70] Ikeda Shigenori, "Yûsei undô to chihô jichitai" in the journal *Yûsei undô*, vol. 2 (1927), no. 1, pp. 4–7.
[71] Ibid.
[72] Ibid.
[73] Tsurumi's article was arranged right next to Ikeda's; see the subsequent pages 8–9. So far I have no way to judge whether Tsurumi ever managed to expound upon his own project, but Tsurumi was one of the *Yûsei undô* members. Ikeda was once a journalist in the *Hôchi shinbun* (*Communication News*), was sent to Germany in 1920 and came back to Japan in 1924. In 1926, he quit his journalist job and established the *Nihon yûsei undô kyôkai*. In his four-year stay in Germany, he came to admire the work and activities of an American eugenicist,

Paul Popenoe, who was in Germany then, and who supported and informed the development of the German sterilization effort based on the California sterilization experience. In the *Yûsei undô*, Ikeda also emulated the German model of physical group training called "Wandervogel" to stimulate a communal sense among the young pupils. See Suzuki Zenji, *Nihon no yûseigaku*, pp. 118–123.

[74] Fujino Yutaka, *Nihon fashizumu to iryô*, pp. 117–118. See also Sumiko Otsubo's chapter 4 "in "Eugenics in Imperial Japan" for a much more detailed process of Marriage Law petitions.

[75] Suzuki Zenji, *Nihon no yûseigaku*, pp. 160–164.

[76] The 1940 version of the National Eugenic Bill was recorded in Yoshimasu Shufu, *Yûseigaku* (Eugenics) (Tokyo: Nankôdô, 1961), pp. 180–187. See also Ma Matsubara Yôko's "Nihon ni okeru yûsei seisaku no keisei: Kokumin Yûsei-hô to Yûsei hogo-hô no seiritsu katei no kentô" (Dissertation, Ochanomizu Women's University, 1997) for a detailed analysis of the Bill's legislation process and statutes.

[77] Nagai Hisomu, *Shizenkan yori jinseikan e* (From a view of nature to a view of life) (Tokyo: Jinbun shoin, 1933), pp. 178–233.

[78] Ibid., pp. 252–292.

[79] Ibid., pp. 292–312.

[80] This article was collected in Nagai's book *Dô to shizen* (Way and Nature) (Tokyo: Jinbun shoin, 1936), pp. 298–299.

[81] Teruoka's article "Eiseigaku no shi teki tenkai ni tsuite" (About the unfolding of the history of hygiene), was published in *Yûseigaka* (Eugenics), vol. 2 (1925), no. 10, pp. 24–28.

[82] Narita Ryûichi "Shintai to kôshû eisei" in *Kôza sekaishi*, Rekishi gaku kenkyu kai ed., vol.4 (Tokyo:Tokyo University Press, 1995), pp.375–401.

[83] See Nagayo's autobiography *Shôkô shishi*, included in *Tekijuku to Nagayo Sensai*, Ban Tadayasu ed., (Tokyo: Sôgensha, 1987), pp. 156–157.

[84] Gotô Shimpei, "Kokka eisei genri" in *Kagaku no shisô*, II (Tokyo: Chimaku shobô, 1964), pp. 289–303.

[85] Asahina Ichio wrote about the Chinese eugenics movement in the report "Senman hokushi ryôjô shûi" in *Yûseigaku* (Eugenics), vol. 13 (1936), no. 11, pp. 20–26.

4

Birth Control or Sex Control?—Politics of Knowledge and Reproduction

In China, a conscious women's movement for self-definition and struggle for women's rights and wellbeing did not begin until the turn of the 20th century, coinciding with the outcry of national salvation. Late Qing reformers such as Kang Yuwei and Liang Qichao promoted a series of campaigns for Chinese women's health and education in order to produce a strong nation. They launched attacks on footbinding and organized new societies to convince parents not to allow their sons to marry girls other than those with natural feet. At the same time, they advocated the establishment of a national system of women's education in China. Knowledge and reproduction were tied together in the beginning of the Chinese women's movement as a remedial prescription to save the nation.

Liang Qichao gave four reasons in 1898 for the necessity of Chinese women's education. First, women should take part in productive labor in order to share the financial burden with men and to progressively change the structure of gender inequality. In order to facilitate female participation in productive labor, they needed education. Second, the conventional view that "only the untalented women was virtuous" should be discarded. Learning, for Liang, would enhance women's knowledge of the world, expand their field of vision beyond mere domestic affairs, and enable them to acquire professions. Third, if Chinese mothers acquired basic knowledge and pedagogy, then children would be able to gain the rudiments of learning before the age of ten. Accordingly, women's education was linked to the nation's survival or demise, its strength or weakness, through the education of their children. Fourth, Liang linked the fate of the nation to the ideas of "preserving the nation," "preserving the race," and "preserving Confucian teachings." In order to secure these spiritual bonds of the nation, men and especially women all needed education.[1] Evidently, it was prenatal care and teaching Liang was advocating. For him, what was needed was maternal training rather than women's advanced education. Even though Liang did not intend and expect to change fundamentally women's role and status, the new reforms led to the development of a recognizable women's movement.

Liang's eugenic linkage of Chinese women's health and education with national survival was further extended by the Japanese woman activist Shimoda Utako (1854–1936), who was devoted to the overseas education of Chinese women students and fostered their revolutionary consciousness and support for popular rights.[2] In 1902, Shimoda bound together the education of both Chinese and Japanese women, to the survival of Asia and the Yellow race in the face of imperialist encroachment from Whites. Shimoda believed that men's and women's education had to be enforced simultaneously in a world of the survival of the fittest and the competition of national endurance. She saw Chinese women's bound feet not only as the insulting symbol of female subordination, but also as an evil institution weakening the female body and passing on its influence to the male progeny. In the age of racial competition, a population of frail female bodies would inevitably reproduce a weak nation, as seen in the poignant contrast of the European White dominance over the Asians. Shimoda argued that women's education be the root of men's education, without which there would be no prosperity, and which concerned not only the wellbeing of the Chinese but also that of the Japanese. Interestingly, within the framework of the Yellows confronting the Whites, Shimoda considered Japan as the leader among the Asians. She therefore concluded that education for Chinese and Japanese women was significant and that Japan should start such a program.[3]

The initiation of the women's movements in Japan and China was embedded in the greater agenda of national survival. The agenda was evoked throughout different stages of the women's movement, and led to a procrastination of an inherent women's agenda for gender equality, independence and universal suffrage.[4] In a sense, the twin expectations of "Good Wife, Wise Mother" (J. *ryôsai kenbo,* or C. *liangqi xianmu*) was a variation and continuation of the national agenda to enlist thoroughly women's productivity and reproductivity, while women were prohibited from participating in decision-making concerning the use of their bodies.[5] Paraphrasing Kang Youwei's words, women became the surrogate wombs of the nation.[6] Women's problems as such posed not only the structural difficulties of inequality, but also the fundamental issue of selfhood. It would be very important to observe and learn how each woman constitutes her self-identity in the process of fighting against these structural barriers.

In the wake of the women's movement, the introduction of eugenics into China and Japan as a form of technology was another variation and continuation of the national agenda, which pledged to provide women proper prenatal care and means of birth control while shouldering them with an ever-growing burden of national survival. Inspired by Margaret Sanger's promotion of birth control in the 1920's and 1930's, Japanese and Chinese women's concern with their own bodies was the only arena where they transcended national/patriarchal boundaries and looked into women's problems cross-nationally. Intriguingly and ironically, Chinese and Japanese women activists and physicians tried to conform with the nationalist pedagogy of serving the country, and in their performative way of evading the biological/national obligation of motherhood they were actually destabilizing and challenging the meaning of nationalist pedagogy. This chapter focuses on women's contestation and negotiation for their own freedom and the right to control their bodies in conjunction with the eugenics movement. With a thorough contextual analysis of the "biography" genre of two women physicians, Yoshioka Yayoi (1871–1959) and Marion Yang (Yang Chongrui, 1891–1983), we observe their embodiments of practice at the performative level, and their positions toward the gendered biological function and the issue of sexuality versus morality, as different approaches to the constitution of their self-identities.

Departing from current scholarly emphasis on women's participation in the war activities, the first section starts with a converging moment in the transnational coalition of women's movements and their ensuing separate developments with local agendas. The second part focuses on biographical analysis of these female scientists at the intersection of the birth-control movement and the eugenics movement. The third section concludes with their different definitions of morality and their strategy of reproduction in contrast with the conventional approaches.

SEXUALITY VERSUS MORALITY: MARGARET SANGER AND THE BIRTH-CONTROL MOVEMENT IN CHINA AND JAPAN

The visit of Margaret Sanger to Japan and China in March and April, 1922, and the ensuing social movement of birth-control campaigns, stirred up issues such as population control, social welfare, women's rights, the prevention of sexually transmitted diseases (STD), and the scientific knowledge of contraception. The birth-control campaigns presented various positive aspects of protecting maternal health, reducing the burden of lower-income families and bettering their children's living opportunities, preventing hereditary diseases, and preventing the unnecessary practice of abortion and infanticide.[7] However, the extensive proliferation of contraceptive technology veiled a threat of unleashing moral constraints on sexuality. This section will examine different reactions to the intricacy of sex, morality and contraceptive technology in China and Japan, and Chinese and Japanese eugenicists' dubious attitudes toward these entangled issues in particular and the birth-control movement in general.

The threat of unleashing moral restraints on sexuality was embedded in the core of newly developed Malthusian doctrines of population control through the incorporation of contraceptive technology. The earlier Malthusian theory recommended "rational, voluntary, humane, and prudential" moral restraint, namely, marriage postponement through the practice of virtuous love and rational control of passions, as an effective preventive check to population growth to prevent all kinds of vice and misery.[8] However, the Neo-Malthusian birth-control movement removed the need for such moral restraint, and advocated early marriage and scientific contraception, which would satisfy sexual need and at the same time prevent any related STD.

For conservative moral theorists, the most threatening element was Margaret Sanger herself, who originally started out from a vehemently feminist position in her coinage and defense of "birth control." In her 1920 book, *Women and the New Race*, Sanger declared, "The basic freedom of the world is woman's freedom. A free race cannot be born of slave mothers. A woman enchained cannot choose but give a measure of that bondage to her sons and daughters. No woman can call herself free who does not own and control her body. No woman can call herself free until she can choose consciously whether she will or will not be a mother."[9] She also encouraged women's free expression of sexuality through access to birth-control technology. She strove to liberate women's sexual pleasure from the patriarchal confinement of marriage, maternity and the accepted Christian morality. She repudiated the moral issue raised by the opponents of birth control, and condemned sexual morals for women as one-sided and repressive in an aim to keep women as an asset to the church, state and men. She sought to create a morality both practical and idealistic for women's own needs, by giving women the opportunity to know themselves and by allowing them to explore their inner natures. The rising new morality, as Sanger defined it, was the feminine spirit animated by joyous and triumphant love; the accompanying free womanhood observed and complied in the best way with the inner demands of its being.[10] In order to achieve such a realization, the primary condition was to obtain adequate knowledge of woman's own sexual nature, in both its physiology and its spiritual significance. "She must not only know her own body, its care and its needs, but she must know the power of the sex force, its use, its abuse, as well as how to direct it for the benefit of the race."[11]

Along the course of development of her program in the American context, Sanger collaborated with the rhetorics of different social groups such as eugenicists, policy makers and socialists in order to gain support for birth-control campaigns, which were soon overshadowed by the gender-neutral goal of family planning and population control. Several studies of the American birth-control movement have shown that Sanger's concern for women's right to control their own reproduction was partly superseded by later concern for the nation's fiscal security and ethnic makeup.[12] However, Sanger's feminist dimension was embraced by the Japanese and Chinese women's-movement activists. Her introduction of scientific contraception provided a technological possibility of reproductive liberty and boosted the fledgling women's movements in Japan and China by means of a scientific basis and rhetorical alignment.

In 1920, Margaret Sanger received a visit from the woman activist Ishimoto Shizue (1897–2001, later surnamed Katô), who talked with her about a visit to Asia. Before the end of the year, Sanger had received an invitation from Ishimoto and the Kaizô group, which had translated four of Sanger's articles on birth control into Japanese in its radical monthly called *Reconstruction* (*Kaizô*). Two months preceding Sanger's March 1922 departure, she found out her visa application was rejected by the Japanese Consul, but she decided to go anyway with the easy approval of a Chinese visa. On the voyage to Asia, Sanger won over her Japanese constituency in the ship, who happened to be delegates to the Washington Peace Conference, and who supported Sanger's entry to Japan. According to Sanger's biographic account, her lectures were well received in both Japan and China as she had given over five hundred interviews in Japan and circulated thousands of copies of the birth-control pamphlet titled *Family Limitation*.[13] She recalled that the China *Times* devoted a whole section to the birth-control subject in its Sunday issue of April 30th and had kept up a steady flow of birth-control articles during her stay.[14]

In the special issue of birth control in the *Lady's Journal* (*Funü zazhi*) in June, 1922, dedicated to Sanger's visit to China, the editorial article first welcomed and celebrated Sanger's mission in China to free motherhood from slavery, and emphasized the unselfishness of reproductive freedom and free love as a foundation from which to build up a new future for better offspring.[15] One author, Se Lu (real name Chen Youqin, 1902–?, a woman writer, high-school teacher and poetry critic) in the same special issue sided with Sanger's feminist position and defined as moral the use of scientific contraception in lieu of infanticide and abortion. She argued that Chinese women, despite their recent achievement in economic independence and educational opportunities, would never be thoroughly liberated without being spared the burden of reproduction, since they would exhaust every bit of mental resource and physical strength from childbearing and childrearing.[16] Another author pushed even further to break down the conventional Chinese belief of childbearing and childrearing as the essence of marriage, seeing them rather as manacles enchaining Chinese women. The author concluded that love is the only element in marriage, and any imposition of childbirth and childrearing would be an overemphasis of biological function without looking at the true nature of marriage.[17]

Translated articles from the writing of Japanese feminists and birth-control advocates echoed the support of women's reproductive liberty. The piece by Ishimoto Shizue offered her own interpretation of birth control and motherhood. She contended that the motivation for birth control was out of true love of children, and that women should seek both mental and physical preparations for childbirth and sufficient care, which in turn underlined true motherhood. In his article entitled "Birth Control and Morality and Religion," the socialist Abe Isoo (1865–1949) reiterated the moral purpose of marriage as a mental and physical union of man and woman, which had nothing to do with childbirth. Based on the supposed evolutionary principle that the higher the organism the lower its fertility, he affirmed that birth control was a manifestation of God's will and in that sense it conformed to the moral purpose of the universe. Another feminist and socialist writer, Yamakawa Kikue (1890–1980), espoused a similar position of husband-wife union as the right goal of marriage. Nonetheless, as a Marxist she rebuffed the Malthusian connection of poverty and overpopulation, and transformed the Neo-Malthusian significance of birth control into a liberating tool for proletarian women to free themselves from the confinements of conventional morality.[18]

Other Japanese feminist expressions of the birth-control idea could be seen in the arena of women's magazines and birth-control journals. For instance, *Shufu no tomo* (Friends of the Housewife) presented forthright commercial advertisements of imported or domestic-manufactured contraceptive devices such as "Sari," as well as "Sana—spermicide cream," a German product patented and certified by seven physicians who claimed to prevent STD, and "Seimori (sex care)—STD vaccine," a vaginal suppository. Other women's magazines like *Fujin sekai* (Women's World) and *Fujin gahô* (Women Pictorial) featured regular posts of "Seimori" and of "Getsuyaku," a *de facto* abortion medication, a name list of private clinic specialists, and ongoing discussion of other contraceptive technologies developed at home and overseas.[19] In view of a tightening censorship, these commercial birth-control advertisements in the guise of STD vaccines were the only medium to survive the State's thought control. These graphic and informative announcements provided several channels for individuals' self-aid and ,expectedly, posed a politically-charged feminist statement of reproductive liberty and self-protection. In addition to existing regional and national birth-control leagues, the Japanese feminists Ishimoto Shizue and Hiratsuka Raichô (1886–1971) set up their own women's league of birth control, which was joined by roughly 1,000 Japanese women, in order to update contraceptive information and distribute the sale of condoms and pessary diaphragms.[20] These discursive alignments of women across national and class boundaries were significant as a challenge to the old regime of patriarchy, and revealed an exciting moment of international coalition of women's body politics.

Interestingly, these birth-control advocates were somewhat bent on endorsing a eugenic pedagogy to conform to the national and racial interest of population quality. For instance, Yamakawa Kikue supported neo-Malthusianism from her socialist perspective, and commented that over-breeding among the proletariat led to weakening of their social standing and thus exploitation by the bourgeoisie. Until the idea that society would guarantee happiness and sufficient education was actualized, the proletariat should reproduce fewer or no children. Moreover, she argued in 1921 that because the prosperity of a race depended not on the quantity but on the quality of its population, over-reproduction should be avoided.[21]

Hiratsuka Raichô, before Margaret Sanger's visit, had incorporated eugenic ideas and defined Japanese women's obligations as being to secure a prosperous *shuzoku* (race) and good *sitsu* (quality) of offspring. From 1919, she devoted herself to the anti-V.D. legislation campaigns, which did not become effective until she cooperated with eugenicists such as Nagai Hisomu and Yamanouchi Shigeo (1876–1973) in the mid 1920s.[22] Her essay "Sanji seigen no mondai" (The problem of birth control) in 1926, emphasized the importance of disseminating the eugenics ideology among people. Since she was committed to the protection of motherhood, she was inclined to push a little further to claim that women should contribute to the nation by producing good soldiers.[23]

In addition to Hiratsuka and Yamakawa, another renowned feminist, Yosano Akiko, also strongly endorsed eugenic pedagogy by calling for "fewer and better children through birth control."[24] Ishimoto Shizue also traded her lip service for eugenic ideas in exchange for support for birth-control campaigns. She said that, in order to enhance the cultural life of the human race, it was essential to produce superior offspring through marriage selection of well-qualified women and men.[25] The feminist pedagogical transaction won the Japanese women later, in 1934, a good package deal in that the National Eugenic Bill incorporated a stipulation to mandate a physical check-up before marriage and to prevent marriage by people with venereal diseases. It also obligated the State to institute a healthcare plan for the protection of motherhood.

However, the persistent feminist expressions of free womanhood at the performative level, and the proliferation of unregulated contraceptive knowledge, invited more suspicion from the eugenicists. The Chinese eugenicist Pan Guangdan observed the development of birth-control campaigns and commented that these campaigns were motivated out of compassionate sentiment for suffering mothers. As such, they appealed predominantly to women and thereby feminists, making the nature of the birth-control movement extremely sentimental. In reality, he felt it was questionable whether the birth-control campaigns ministered to a practical need of Chinese society. Like his American mentor Charles Davenport, Pan was reserved about the birth-control movement, since contraceptive technology had been abused by American middle and upper classes for sexual gratification. He considered birth control a double-edged blade that would render a population out of proportion to its class distribution, and thus would extinguish the upper and middle classes in the long run. He suggested that birth control be selectively applied to the lower class, which could not afford better upbringing, and in which women suffered over-reproduction and physical fragility. Facing the vast majority of Chinese peasants, Pan was conscious of his class definition. Different from the conventional criterion of wealth and economic status for class division, Pan adopted the idea of "civic worth" originating from British eugenics, which emphasized, in addition to economic measures of livelihood, creativity for the public good.[26]

The Japanese eugenicist Nagai Hisomu's opposition was even more radical. He declared, "Birth control is racial (or national) suicide (*minzoku jisatsu*)." He commented that the earth's resources, in terms of food, could feed a world population of five to nine billion; if they lived lavishly like Americans, maybe 2.3 billion, if moderately like Germans, 5.6 billion, and if frugally like Japanese, 22 billion. Since the world population was 1.8 billion at the time, there was much room for population growth. Moreover, Nagai was confident that scientific technology would definitely enhance productivity in the future. In view of international politics and competition, in the ideal condition of peaceful competition the final victory would belong only to the race that could endure and prevail. To reduce lopsidedly the biological power of one's race, Nagai remarked, would be like a fool strangling himself.[27]

Despite the eugenicists' suspicions and admonitions, subsequent debates on sexuality versus morality were energetically pursued as a hot topic in the women's magazines and other journals. In China, these controversies led to the 1925 publication of a collected volume entitled *Xin xing daode taolunji* (New Sexual Morality), which was compiled by Zhang Xichen (1889–?) and assisted by Zhou Jianren. Zhang was once the editor of the *Lady's Journal* sponsored by the Commercial Press, and was criticized harshly for his unreserved editorial approach to the perceived "indecent" subject of sexuality and contraceptive knowledge. Zhang hence resigned as editor in 1926, and established a new publishing house, Kaiming Shudian. Here he launched his own magazine, *Xin Nüxing* (New Women),[28] which was devoted to the feminist issues of sexuality and Chinese women's self-knowledge.

In the *Xin xing daode* discussion volume, the articles sought either to define and redefine the moral dimension of sexual desire, or to explore sexual repression in the conventional and new moralities. The contextual meaning of these statements could be seen as a manifestation of Sanger's moral principle of birth control encountering local situations. Zhang and Zhou, the two main contributors, first stated their definition of the new sexual morality and its scientific basis. Zhang saw continuity—rather than the conventional binary separations of moral/immoral, monogamy/polygamy, altruism/self-benefit—in order to neutralize popular perceptions of sexual desire and gratification. Following August Forel's moral law, Zhang gave his judgment of morality as an equilibrial and harmonious combination of altruism and self-benefit. In order to elaborate this definition, he gave an idealistic eugenic example of a healthy and bright man and woman falling in love with each other, marrying, giving birth to and raising healthy offspring with good care and education. Anything other than this principle would be considered immoral to different degrees.[29]

With an evolutionary overview of cultural and temporal differences, Zhou presented two scientific criteria for sexual morality. First, human desire is a natural norm, and yet the result of desire might harm oneself and others. Second, the results of sexual behavior are connected to national interest and the ethnic future. Chastity, concubinage and prostitution were regarded as lopsided woman-subjugation. Divorce freedom, like free love, should be taken as a private matter and hence be made subject to one's personal decision. What Zhang and Zhou were proposing was to consider love affairs and courtship as private matters with which only the two parties were directly involved, and to separate them from the public concerns of the formation of family and consequently of society. Zhang and Zhou's statements elicited criticism from Chen Bainian, a philosophy professor at Peking University. Chen Bainian responded and argued that Zhang and Zhou's proposition in turn justified the unjust social phenomena of chastity, concubinage and prostitution by using the dichotomy of private/public as a pretext to withhold them from social interference.[30] As the discussion went back and forth like a tug of war, the definition of sexual morality was boiled down to self-restrained sexual control as opposed to excessive indulgence.

Lu Xun once commented that Zhang and Zhou's discussion on sexual morality was ahead of its time and hence was not well received by the public.[31] However, in the eyes of some Chinese women, the feminist expressions of both Zhang and Zhou were actually inadequate. Their indiscriminate and gender-neutral definition of sexual morality, in particular, alienated some Chinese women readers and writers. They were unsatisfied with such a repressive discursive result at a time when, on the one hand, the inception of sexual enlightenment had begun to awaken Chinese women's physical sensations and assist their self-exploration; and on the other hand, Chinese men had excessively indulged themselves and aggravated social evils to women such as chastity, footbinding, concubinage and prostitution. One anonymous woman reader addressed the debate and commented that conventional polygamy and concubinage were rather men's showy desire for power, authority and wealth, and that women's sexual desire was never in the picture or a subject of knowledge. She further remarked that the moral criterion of sex should be based on whether there is love between the two autonomous parties. Therefore, issues of polygamy, monogamy, and extramarital affairs were irrelevant, and mutual chastity would develop consequently when love endures.[32] Obviously, not only was women's sexual desire never satisfied, but women's desire for sexual knowledge was forever repressed.

This discontent flamed an attack on the *Xin Nüxing* (New Women) journal and led to the publication of a new journal *Xin Wenhua* (New Culture), which was aimed at disseminating sexual knowledge and aesthetics, and promoting provocative discussion on women-centered issues such as Chinese women's right to property succession. Its first issue included a critique of *Xin Nüxing*, from a female reader pen-named Lady Lina. It contended that this journal no longer lived up to the feminist expectation of providing a discursive space for issues such as women, family, children and sexual desire.[33] Unlike the banal subjects and idle discussion in *Xin Nüxing*, the chief editor of *Xin Wenhua*, Zhang Jingsheng (1888–1933),[34] declared in the first issue that this new journal was determined to reconstruct a new media culture that would peel off any masquerade of repressive morality. Rather, it would pursue the forbidden topics underneath, especially with respect to sexual education and aesthetics.[35]

In the same issue, Zhang provided his expertise for exercising sexual desire and acquiring its best benefits. He advised against any use of aphrodisiacs and suggested the approach of physical training as used in the internal part of Daoist alchemy, such as exercises of upper and lower abdominal breathing, and penile and vaginal suction. These exercise skills, like martial arts, would not only increase blood circulation and enliven sexual organs, but also augment physical and spiritual strength, which could prolong the length of lovemaking and satisfy one's female partner. In order to intensify one's sexual desire, Zhang recommended two systems of sexual game, one the multi-lovers system (*qingren zhi*) for mere sexual fantasy, and another the system of physical intercourse (*jiaogou zhi*) confined within the monogamous relationship. Zhang believed that these two creative systems would reconcile the dichotomy of sexuality versus morality, on the one hand guaranteeing sexual gratification and on the other hand monitoring sexual control.[36]

Zhang's provocative statement on sex, and the publication of his personal sexual history, invited more attacks on his astounding assertion. In particular, he claimed in the first volume of *Sexual Histories* (*Xing shi*) in 1926 that perfect conception between a sperm and an ovum would develop a better embryo and thereby a healthy child, when the ovum was freshly produced and enlivened by secretions from the vaginal passage, which he named the third kind of fluid (*disanzhong shui*) in addition to the first kind from the labia and the second from the clitoris.[37] Zhang believed that children conceived in the moment when their mothers excreted fresh eggs and the parents achieved orgasm in their intercourse, would be healthier and more intelligent. Zhang defined this eugenic connection with perfect lovemaking as an aesthetic sexual desire. He further elaborated that the discharge of alkaline fluid, when women reached their orgasm, would facilitate the activities of vigorous sperm in making their way to the uterus. In addition, vaginal secretion would speed up the maturation process of the egg cell, and preserve it for a longer length of time. The spasm in the orgasm would expedite the ovum's descent through a fallopian tube in order to meet with the sperm.[38]

Zhang diagnosed the feebleness of the Chinese race, in comparison with Anglo-Americans, was due half to postnatal factors, and half to the prenatal determinant that women were unable to discharge the precious third fluid, making the ova stale and the sperm worn out by sterile vaginal acid at the time of conception. Europeans were able to concentrate on their lovemaking and most women were able to discharge vaginal fluid, and hence they themselves were much more lively spirited and their children were far healthier and stronger than their Chinese counterparts. Zhang argued that the conventional eugenic emphasis on the parents' character and morality was not at all reliable, because superior parents could not guarantee good offspring. Zhang suggested that only the perfect conception of fresh ovum and strong sperm is able to produce a good healthy child. Zhang thus boiled down his definition of eugenics from the perfect conception of fertilized egg, and perfect sex, to the third fluid. Subsequently, he discerned two reasons why Chinese women were not able to discharge the third fluid. First, Chinese women were shy about sexual expression, and were repressed by the conventional moral yardsticks that considered women's sexual interest as a sign of debauchery and thereby inappropriate for the roles for a good wife and wise mother. Zhang speculated that many women's diseases and premature deaths were caused mainly from sexual repression, in addition to the abuse they received from their mothers-in-law, which was another indication of sexual repression and which formed a vicious cycle. Second, Chinese men and women—but especially men—did not know how to make love. For men, sex, was generally merely functional ejaculation regardless of the female partners' sexual gratification. Speaking like an experienced sex therapist, Zhang provided a series of ways in which men could electrify women's sexual desires. He encouraged women to play an active part as well, and help themselves reach the climax of extreme excitement in order to discharge the third fluid.[39]

Zhang's seditious statements infuriated two differently-positioned eugenicists, Zhou Jianren and Pan Guangdan. Zhou, as one of the editors of the *Xin Nüxing*, denounced Zhang's Daoist esoteric exercise of sexual skills as feudal residue, and chided Zhang for instilling new youth with misguided information. He also questioned the purpose of penile and vaginal suction in Zhang's assertion. Zhang repudiated Zhou as a hypocrite who not only was dishonest to his own sexual desires and did not take sexual intercourse seriously, but also rashly accused others' efforts in circulating such valuable knowledge of spiritual and physical wellbeing.[40] In contrast, Pan Guangdan was extremely upset by Zhang's making an equation between eugenics and women's vaginal secretion, and debunked it as pornographic and scientifically baseless. Whereas Zhang placed both sexual enjoyment and reproduction in a unified function of human sexual organs, Pan tried to differentiate these two as disparate capacities without any necessary interconnection in their activities.[41]

Pan enumerated his criticisms. First, vaginal fluid, hardly a novel idea in physiology, was secretion from the glands of *Bartholin* to lubricate the vulva during sexual excitation. At the point of maximum secretion, the woman would experience voluptuous feeling analogous to the man's orgasm. Without any statistical data, Zhang's assertion of Chinese women being unable to discharge vaginal fluid was completely speculative. Pan suspected that sensational instigation of the public's sexual appetites might be the selling point at which Zhang was targeting his pronouncements. Second, in order to justify orgasm as irrelevant to conception, Pan cited August Forel's account, "the neck of womb, during the venereal orgasm of the woman, executes movements of suction in the glans penis...but it is certain that the female orgasm is useless for conception. Absolutely cold women, incapable of the least voluptuous sensation are as fruitful as those who have pronounced venereal orgasms."[42] Third, Zhang's speculation on the connection between vaginal fluid and faster ovulation was absurd and meaningless. Fourth, the only actual connection between ovulation and sexual intercourse would have to do with the possibility of simultaneous fertilization of two or more eggs, thereby giving rise to the formation of fraternal or identical twins (which is not true). Pan provided Charles Davenport's statistics on identical twins to affirm its role in heredity, instead of the unsubstantiated factor of vaginal fluid. Fifth, Pan again quoted Francis Galton's statistical work *Hereditary Genius* to support the conviction of familial heritability of talent or intelligence. Pan concluded that Zhang's salacious reduction of eugenics to vaginal fluid was purely nonsense or ignorance.

Immediately after disclaiming Zhang's eugenic definition in May, 1927, Pan wrote another article in June, delving into the professional qualification of public figures commenting or circulating their knowledge on sexual problems in the mass media. He was insinuating that Zhang was traumatized by his own sexual frustration, which disqualified him in the first place. As a philosophy professor at Peking University, Zhang was not properly trained as a sex therapist. Pan further critiqued Zhang's motivation, claiming that such a sensational topic arose out of his own hunger for fame instead of his conscience for social morality. Zhang's compilation of various personal accounts of sexual experiences into a volume of sex histories was not a rigorous academic work, and hence subscribed to unwarranted folk beliefs.[43]

From Pan's and Zhou's vehement critiques of Zhang's eugenic reduction, one could discern a spectrum of different positionings in relation to the issues of sexual pleasure, reproduction and racial betterment. For Zhang, these three objectives came together as an interlinked process and should not be separated. For Zhou, sexual pleasure was a private matter and should be an issue merely for the two parties concerned, and was not directly connected to the public domain unless it resulted in pregnancy and the formation of family, whereby it became part of the public concern for population quality. For Pan, sex had two disparate functions, sexual pleasure and reproductivity, and these two should be dealt with by different approaches. For Pan, sexual pleasure seemed to be a byproduct in a wide range of performance of the sexual organs and hence should not override and take over the major avenue of reproduction. It would not be a farfetched assessment that Pan's perception of reproductivity as an inherent biological function was equivalent to a sexual morality, whose focus was on the ability to reproduce better progeny, rather than on issues of sexual pleasure.

When Pan translated Havelock Ellis's essay "Sex Morality" into Chinese in 1934, he followed Ellis's eugenic view that sex should be finally aimed at the reproduction of progeny. In his preface to this translation, he added the Confucian principle of sexual morality, *jie*, with the annotation of "*Fahuqing, zhihuli* (When feelings or emotions blossom, they rest upon propriety)." The Confucian technology of self-cultivation, emphasizing spontaneous self-adjustment of appropriate timing and positioning, should also be applied to the matter of sex, which did not concern the public/private distinction, but rather the person's selfhood. The emotions in action are self-motivated and hence self-regulated. They start when/where they should start, they rest when/where they should rest. While these self-regulated emotions are the manifestation of virtue in one's self, they in turn constitute one's selfhood.[44] Pan's tautology of self-extension and cybernetic return of self assumes the perfect functioning of a mature selfhood, which seemed to be male-gendered and not to apply for women. He considered that before women could self-determine and take responsibility for their own sexual behavior and procreation, they should have the qualifications of economic independence, and the education of professional training and sexual hygiene. With these qualifications, a woman could be the master of herself.[45] However, in the Chinese population of the time, how many women could be qualified as being the masters of themselves? How could a woman not end up being guided and protected by her copulative master, who had better self-control? In other words, the sexuality of the majority of Chinese women, in Pan's assessment, was in the stage of immature selfhood and had not reached self-regulated status. What about the qualified minority who could be the masters of themselves?

Fearing that these self-mastered women might think and do otherwise, Pan suggested that they first devote their first or more decades after marriage to motherhood, leaving until later their careers as a reconciliation of knowledge and reproduction.[46] However, among this minority of capable women, there were a small crowd of women activists advocating free womanhood and gender equality. Pan commented that these women leaders were knowledgeable and talented, but mostly indifferent to motherhood. He subsequently criticized the flaws of the women's movement to be the exaggeration of the minority's problems as the majority's, and the imposition of the minority's solution on the majority, who were themselves content within the family structure and their role in motherhood.[47] Evidently, Pan was disturbed by the split between knowledge and reproduction in more recent developments, as these two had gone hand in hand in the initial formation of the women's movement.

The Japanese eugenicist Nagai Hisomu also expressed a similar critique of the women's movement, whose goal of full scale liberation and equality did not take into consideration gender differences and women's divine mission (*tenshoku*) of reproduction. Nagai advised strongly that women should maintain full self-awareness of love (*ren'ai*) as the fundamental instinct for species survival. Women should build on the foundation of their own inborn organic system, and its reproductive function, in order to come up with their own demand for rights in the face of eugenic problems. He pinpointed as a social crisis in Japan that the mortality rate was rising higher than in European countries, while the birthrate was progressively declining. The crisis was even more aggravated by the popularization of hedonistic contraception, imported and masquerading under the beautified title "birth control." He exhorted Japanese women to cautiously examine this crisis from their own perspective, reveal the masquerade of birth control and cast off the spell endangering the nation. Interestingly, he marked the realization of sex difference and respect for reproduction as the first step of women's self-awareness, marching towards the aim of reconstructing a foundation of gender equality in the aspects of law, politics, education, religion and social life.[48]

Abhorred by the eugenicists, the birth-control movement proved to be a middle- class-centered phenomenon. According to the 1934 report of Marion Yang (Yang Chongrui, 1891–1983), the Peking Committee on Maternal Health helped in the treatment of 99 cases, 96 for birth control and 3 for sterilization. Among the 99 cases, 98 women were from the middle and upper classes.[49] Moreover, based on the analysis of 1273 cases handled by the Peking Committee on Maternal Health from March, 1930 to June, 1936, the pioneer scholar Su Ru-chiang (her thesis *Birth Control in China* won her an M.A. degree from the Department of Sociology, University of Chicago, in 1946) concluded that the work of the Committee was limited to the urban district and largely available only to the middle and upper classes of the urban population throughout its three decades. And the effect of the birth-control campaigns did not penetrate deep into the peasantry and the wide range of rural areas.[50] According to the documentary entitled *Chinese Women in the 20th Century*, the Rockefeller Foundation had supported Chinese birth-control campaigns by supplying condoms and having the Peking Union Medical College manufacture contraception pills. However, the idea of using condoms did not seep into Chinese men's mentality, and the pills were only good for winter, since they dissolved in summer when the use of iceboxes was limited to certain urban areas.[51] This technological difficulty partially explained the birth-control campaigns as an urban phenomenon in China.

For Japan, contraception technology introduced by Margaret Sanger was originally aimed at American working-class women, but there were no comparable sanitary facilities in the Japanese household setting. The easily accessible douche in the United State was expensive and hence unaffordable for the Japanese working class. Condoms purchased by mail order were censored. Some of the Japanese medical practitioners complained that in practice these contraceptive methods did not work as smoothly as expected, since inadequate douche sanitation and reused condoms and pessaries would only increase risks of infection.[52] The proletarian biologist Yamamoto Senji, although an ardent advocate of the birth-control movement, criticized Sanger's campaigns in Japan as being monopolized by the elite and medical experts, and hence providing service aimed only at the elite and the upper and middle classes.[53] Accordingly, these developments of the birth-control campaigns and the bifurcating outgrowth of reproduction knowledge plunged the eugenicists into unfavorable judgment on excessive sexual indulgence and individualistic selfishness among the middle class.

POLITICS OF REPRODUCTION AND KNOWLEDGE: WOMEN SCIENTISTS AND THEIR LIVES

In view of the similar outlook on the parts of both Chinese and Japanese eugenicists towards female sexuality, and the similar approach from their professional authority to curtail women's pleasure-oriented pursuit of sexual knowledge, it is intriguing to observe the positioning of Japanese and Chinese women gynecologists, who were the masters of this particular set of knowledge at the levels of eugenic pedagogy and performative embodiment in their life experiences and social endeavors. The two particular gynecologists I am examining here are Yoshioka Yayoi and Marion Yang, who were not less eminent social actors in the eugenics movement. Both of them worked all their lives with eugenicists, hygienists and sociologists for the cause of national interest—the project of public health—and were marveled at by their colleagues and compatriots as moral examples for their spirit of sacrifice and selflessness. Different from their feminist counterparts who merely endorsed the eugenicists' pedagogy, these two gynecologists were true believers and served their lifetime for this particular cause. A suitable trope to characterize their relationship to eugenic ideology, would be that they were the eugenicists' chaste lovers who shared the same dream even though sleeping in different beds (*tongmeng yichuang*). In contrast, their feminist counterparts might have been said to sleep in the same bed with the eugenicists, yet dream different dreams (*tongchuang yimeng*).

The biographical materials I am examining are either from their own autobiographies, or from memoirs and personal accounts written by friends, disciples, family members and colleagues. Autobiography as opposed to biography does not take up a uniquely privileged position of superior authority for the subject "I," since the moment the author narrates "I" she/he steps out consciously from her/him-self in a moment of recollection or reflection.[54] The autobiographical author and other biography writers all share a common subject—the memorable name that is uttered over and over again, in the aim to hold together and attach meaning to every fragmented piece of memory. They are manifold voices and yet seem to orchestrate a chorus that renders a formation of subjecthood, and from the chorus I would like to tease out their performative embodiment of what constituted womanhood in the controversy of sexuality versus conventional morality.

An explicit comparison of Yang's *Fuyou weisheng* (Women and child hygiene) and Yoshioka's *Anzan to ikuji* (Safe delivery and childrearing) textbooks on maternal health and child care—one emphasizing and the other omitting "birth control"—plunges us into an inquiry of their different performativeness in the convergence of the birth-control and eugenics movements. Let me first provide some snapshots of Yoshioka Yayoi's and Marion Yang's public activities. Yoshioka was born in Shizuoka prefecture. In the Meiji era, she was one of the very few women doctors. She graduated from Saisei Gakusha Medical College in 1892 and soon opened up her own hospital. In 1900, she established Tokyo Joi Gakkô (Tokyo Women Doctors' School), and became the president and director of the hospital attached to the same school. Based on the promulgation of *Senmon Gakkô rei* (College Decree), she was appointed as the president of Tokyo Joshi Igaku Senmon Gakkô (Tokyo Women's Medical College) in 1912. In 1923, she became the president of Shiseikai, an organization composed of alumni of Tokyo Women's Medical College, and the director of Shiseikai Hospital.

From the 1930's, Yoshioka joined the Racial Hygiene Association in Tokyo.[55] Not only did her only son Yoshioka Hiroto and her talented disciple Takeuchi Shigeyo (1881–?, the first Ph.D. and M.D. graduated from Tokyo Joshi Igaku Senmon Gakkô, where Yoshioka served then as the president) join the Racial Hygiene Association and often contribute research articles to the Association Journal *Minzoku eisei*, but also the entire school Tokyo Joshi Igaku Senmon Gakkô joined as an Association member. During this period, she also served as the president of Nihon Joikai (Japan Women Doctors' Society) and

Dai Tôa Seikatsu Kyôkai (Great Asia Livelihood Association), a counselor of Dai Nippon Fujinkai (Great Japan Women's Society) and a technical officer in the Ministry of Health and Welfare. She was purged from public posts immediately after the end of WWII in 1946. In 1951, she was rehabilitated and became the president of Tokyo Women's Medical University. She was awarded the Fujin Bunka shô (Women's Cultural Prize) and Kun Shitô Hôkan shô (The Fourth Order of the Sacred Crown) in 1955. Other posts included the supreme counselor of Nihon Joshi Shakai Kyôikukai (Japanese Women's Social Education Society) and the president of Fujin Kôseikai (Women's Welfare Society).[56]

As for Marion Yang (Yang Chongrui), she was born in Tong county, Hebei province, in 1891. In 1917 she received her M.D. from the Peking Union Medical College for Women, upon which she worked in the gynecology and pediatrics wards in a small municipal hospital in Shandong province. In 1921, she returned to her *alma mater* and worked as a residential intern. In the same year, when the Peking Union Medical College set up a department of Public Health and Preventive Medicine, the Department head John B. Grant encouraged Yang to join a survey project investigating the cause of the high mortality rate of infants and women right after delivery. In 1925, she received a fellowship to pursue advanced study in gynecology at Johns Hopkins University for a year. Before she returned to PUMC, she did an inspection tour of public-health work in America and Europe.

From 1927, she was devoted to the work of maternal care and child welfare, and took up public posts related to this cause. From 1929, she was appointed as the president of the First National Midwifery School, where she once recruited Chinese eugenicist Pan Guangdan to teach the subjects of sociology, psychology and eugenics. In 1930, she also joined the projects of mass education and rural reconstruction with James Yen (Yan Yangchu, 1894–1990) and Liang Shuming (1893–1988). In order to improve the quality of maternal health and childcare, Yang also performed birth-control work both in clinics and through propaganda. In 1932, she also served as an editor of the population column in the Morning Post (*Chen Bao*) and Chinese Daily (*Zhongwen Ribao*) in Beijing. In 1936, Yang took charge of the arrangement of Margaret Sanger's second visit and her lectures and activities, for which she received virulent attacks from opponents. During the wartime period (1937–45), she worked as a technical officer in the Ministry of Health and Hygiene, and was a delegated expert to the World Health Organization in the United Nations.

When the Chinese Communists took over China in 1949, she was working as the vice-director of the Maternal Care and Child Health Division in the United Nations. Upon the invitation of Li Dequan (1896–1972), then the minister of Health and Hygiene, she returned to China and served as the deputy director of the Bureau of Maternal Health and Child Welfare until the anti-Rightist Movement in 1957, when she was purged for the accused "crime" of her promotion of the 1930's birth-control campaigns. After the end of Cultural Revolution in 1979, Yang was vindicated from the "Rightist" charge and the purge was reversed. However, her prior post of deputy director was filled and she worked as one of the counselors to the Bureau.[57] It is important to note that the resurgence of Marion Yang in biographic form, and the recognition of her contribution to the birth-control movement and midwifery training, has to do with the redirection of the eugenics campaigns and the State's population policy starting in the 1980's. This redirection made possible the rehabilitation of people such as Yang and the eugenicist Pan Guangdan, who were previously charged as "Rightists" and purged from their administrative and academic posts.[58]

In her autobiography, Yang specified the biological fact of being a woman as a reason why she was devoted to gynecology and maternal and child health as a lifetime cause.[59] Despite the fact that feminine speech was not particularly visible in her writing, the biological fact of being a woman was never a source of discontent, complaint or torments. Instead, it was a source of inspiration, identification and justification to claim the turf of knowledge of the female body. Yang's elder brother was a surgeon, and Yang was treated equally in the family and received her education as well. Her father was the Chinese language and history teacher at the Union College, and probably because of this connection she became a student in the Chemistry and Physics department at the Union Medical College for two years. Upon finishing the preparatory program, she entered the Peking Union Medical College for Women and received her M.D. there. Ever since her college life, she had participated enthusiastically in YMCA social work and charity activities. She might have been a Christian, but she did not mention any of her religious background in this autobiographic piece, which was written in October, 1949, after the Chinese Communist Party took over China and Yang was invited back by the Communist regime.

Li Dequan, a renowned feminist activist, the vice-chair of the National Women Organization and once the minister of Health and Hygiene in the People's Republic of China, presented Yang as a celestial savior to shape the destinies of a multitude of women and children. She wrote that Yang had been devoted to the cause of maternal health and child care ever since her childhood when she saw children and women dying too often in her home town.[60] Li's daughter Feng Lida further presented Yang as a new maternal hero who toppled feudalistic superstition and ignorance and the evil obstetric practices residual from the past that had killed numerous sisters and their offspring.

However, according to Yang's own account, before and after she received her M.D., or even before she did her residency in gynecology in PUMC, she did not particularly feel committed to this field. It was not until

1921, when PUMC set up a department of Public Health and Preventive Medicine and she was recruited by the departmental chair John B. Grant, that she recognized the importance of preventive medicine in prenatal care. She recalled two episodes of psychological impact that drove her toward the work of prenatal care and midwifery training. A pregnant woman about to deliver was sent to PUMC for emergency treatment, but died in front of the entrance door because her uterus had ruptured. Yang lamented that such things could be prevented if there were prenatal check-ups. She subsequently set up a Health Station around that deceased woman's village. Another incident occurred when a villager in Sanhe county wrote to PUMC asking if there was any magic bullet for the safe delivery of mother and child, and the prevalent problem of newborn mortalityand the letter was referred to Yang from Grant. Hence, they organized a survey team in 1924 to Sanhe county and its vicinity to investigate a devastating disease commonly called *Siliufeng* (*tetanus neonatorum*) often found in infants.[61] Xiong Bingzhen has documented systematically traditional Chinese childcare of newborns in the medical texts in which cases of *umbilical tetanus* were found. Conventional preventive measures were a complex of procedures of biting off umbilical cord with a cloth cover and cauterizing it off completely. It is hard to establish a causal connection between high mortality from neonatal tetanus and ineffectiveness of traditional medical practice, because old-styled midwives might not acquire or follow same procedures due to lack of training or limits of material condition. It is important to note that these midwives were often illiterate like their Japanese counterparts. In this survey experience, Marion Yang discovered her own womanhood and its ontological concern—women's security (*an*), danger (*wei*), sickness (*ji*) and suffering (*ku*).

In addition to the highly infectious *tetanus neonatorum*, puerperal fever was another major factor in the high Chinese perinatal mortality. Yang provided comparative statistics for Chinese infant and maternal death rates. Whereas out of every thousand live births in the nation 15 mothers died, the statistics found for Britain were 3, for the United States, 5; and for Japan, 4. As for infant mortality, the rate was as high as 250–300 per thousand, while for Britain it was 90, for the United States it was 87 and for Japan it was 187. Yang explained that such a high mortality rate in China was caused by, first, inadequate midwifery knowledge of pathology and physiology to handle difficult labor; second, poor sanitation and incorrect sterilization procedures; and third, food hygiene and insufficient nutrition for infants and mother after birth.[62]

However, these situations could be rectified immediately if there were appropriately trained midwifery in place. The gynecology department at PUMC suggested that Yang set up a short-term midwifery training program to recruit women with primary-school education and send them across China in order to replace the old generation of reproductive techniques, which in the eyes of modern physicians were backward and life-threatening. Yang had her own reservations about this program. She was concerned about the future of these instantly-trained midwives, who would perform the low-esteemed task of childbirth delivery. She could foresee that once they had performed their instrumental function they would be left behind by the society as it moved on. Yang did not want to appease women's suffering by adding social distress to other women.

For that reason she threw herself into the business of Chinese midwifery education. After her advanced training at the Johns Hopkins University and tour inspection of midwifery education in America and Europe, she presented her blueprint for a professional midwifery school at the seventh annual meeting of the Chinese Medical Association in 1928. (This was at the time when the Nationalist government had unified China and was ready to home in on a series of National Construction projects.) In the design, she proposed two complementary divisions, one for a regular program of two-year requirements, and another for short-term programs of six-month and two-month classes. The former was targeted at high-quality training as a preparation for next-generation midwifery faculty, whence China could reproduce its faculty members and eventually become independent from foreign aid and service. The latter was aimed at producing massively and instantly human resources, by retraining old-style midwives and recruiting any interested women, towards the extensive and pressing need to bring down the high mortality rate of infants and women after delivery. Yang specifically emphasized the facility of a delivery ward attached to the midwifery school for both teaching and clinical service. This proposal was accepted and approved by the Ministry of Health and Hygiene, and the Ministry of Education, and hence Marion Yang delivered the birth of the First National Midwifery School in Beijing in 1929. At this historical juncture, the Nationalist government considered old-style midwives (often called *jiesheng po*, *chanpo*, or *laolao*, literally meaning 'old delivery ladies') between the ages of 30 and 60 to be qualified and began to mandate them to register with the government and enroll in a two-month program on sepsis and the rudiments of modern midwifery.[63] Yang subsequently helped reproduce 54 regional midwifery schools across the country before the outbreak of the second Sino-Japanese War in 1937.

How did Marion Yang get involved in the birth-control campaigns? According to Lei Zhifang, one of the biography contributors, Yang was seriously disturbed by numerous clinical cases in which, due to intensive multiple births, women patients suffered chronic symptoms such as overstretched cervical muscles and ligaments, ruptured uterus or vagina, or *uterus prolapsus*. These women were distressed by intensive over-reproduction and their inability to avoid such a depressing reality. Shocking statistical records prepared by Yang's midwifery ward between 1929 and 1932 showed that women experienced as many as 15 births and started to give birth as early as age 15. These clinical phenomena worried Yang with respect to the mental and physical health of these suffering mothers. In a sense, Yang felt that these women were unnecessarily abused, which would result in a serious cost with respect to inability in childrearing, which would jeopardize the physical well-being of the individual, family and nation. Hence, Yang found birth control to be a key answer to these problems.[64]

Yang's participation in the birth-control campaigns started from 1930. Sponsored by the Peiping (Beijing) Committee on Maternal Health, Yang, together with James Yen and Chen Da (1892–1975, Ph.D. sociology, University of Chicago, renowned demographer), and faculty members from the Sociology Department at Yenching University, and PUMC's Department of Public Health, conducted an intermittent clinic in Beijing to advise on birth-control devices. It was the first time that the birth-control campaigns had materialized on the level of clinical practice. At the same time, Yang served as the editor of the "Population Column" in *Morning Post* to help propagate contraceptive culture and answer readers' questions on related issues. From March of 1930 to February of 1933, Yang reported, the clinic received 159 cases and treated 98 women patients. The clinic primarily advised the use of diaphragms and spermicide, because the IUD was not popular among women for reasons of insufficient confidence and superstition, the condom was not favored by men, and the method of rhythm was not reliable. Yang pointed out the noticeable fact that the majority of patients came from the middle class. For instance, of 98 wives, 16 were college-and 27 middle-school-educated. Of 100 husbands, 55 received college education and only 3 were illiterate, and 90 could be classified into the category of middle-class income.[65]

Historians and sociologists often concluded, based on Yang's official report, that the birth-control movement at this stage was a middle-class phenomenon. The eugenicist Pan Guangdan suspected, and other critics of the birth-control movement also commented, that birth control had became a convenient tool to gratify individualistic sexual desires on the part of the middle class, and thus bypass their social obligation of reproduction. However, Yang's birth-control project did not target merely middle-class women. Instead, it was aimed at any women who came to her for help as "patients" suffering the aforementioned symptoms and other related diseases. One anecdote could footnote her position on the doctor/patient relation. When she was a visiting scholar at the Johns Hopkins University, she brought in a black female patient suffering a hemorrhage symptom of hydatidiform mole, which frightened a lot of her colleagues in the operation room. In spite of their discouraging sarcasm, she successfully operated on this lady and ended her excruciating pain. Yang commented on this incident to her students, "One should not despise working-class women nor discriminate against black women. To relieve their pain is more significant than to learn to do major surgery in the operation room."[66]

Furthermore, her birth-control activities were not exclusively limited to the aforementioned official and public appearances. In her teaching materials prepared for the First Midwifery School curriculum, she specifically added a chapter on "Birth Control," which detailed the needs for birth control and a series of contraceptive measures and procedures. She also provided outpatient services of birth control every Saturday in her regular Maternal Health Station in the Dongcheng District in Beijing. In the regular program of two-year training at the First Midwifery School, all students were required to do their internship in the birth-control outpatient service four times before they received their diplomas and were sent out to every corner of the country to provide clinical service and help train instantly midwifery and maternal health workers at the local level.[67] One can imagine that this resulted in a geometrical expansion of birth-control knowledge and should not underestimate its far-reaching impact on the reproductive culture of rural areas.

The public image of Marion Yang, as a martial warrior of enlightenment bent on eliminating feudal residue and ignorance, impresses us as an emotionless cold statue. However, Yang was pictured privately by her own students as the persona of the benevolent mother who gave birth to and nurtured the new generation of midwifery workers. Despite the fact that Yang never married and never reproduced her own biological offspring, her students acclaimed her life with the verse, "the spring silkworm never stops impregnating silk until its last breath of life, and the harvest of plums and peaches satisfies the world." In Chinese language, spring silkworm, plums and peaches often symbolize fertility and are used on the occasions of love affairs and weddings. But for Yang's specific embodiment, they signify the reproductive power of her knowledge. The signification is not limited to the mere growth of her own knowledge, but also draws an image of bearing seeds of the next generation for reincarnation.

But the use of the symbols "spring silkworm, plums and peaches" for reproduction without sex is very intriguing. One of her students recalled that Yang was well known for her unbound feet, frugal lifestyle and staying single. She was worshipped as an idol by many professional women and was respected by her male colleagues, for her sacrifice of a much more glamorous career in medical practice and for becoming a pioneer in the fields of public health and midwifery. Yang's student entitled this memoir "Her heart was open only to one side" and commented that money, glory, gourmet food and luxurious clothing would never excite Yang's passion, only knowledge and the success of the field. She quoted Yang's words, "I was married to the business of maternal health and childcare."[68] Her intimate union with gynecology indicated the unity of women as object and subject of knowledge. It also testified to a synthesis of her professional identity and self-identity as a woman. Her romanticization of knowledge of the female body characterized her sexuality without sex, as a way of bypassing the assigned biological function of reproduction and arriving at her own destination of the reproduction of knowledge. Her life embodiment of eugenic pedagogy unexpectedly and ironically turned against the eugenicists' agenda that women like Yang should have reproduced as many progeny as possible.

Both Yang and Yoshioka conformed to the traditional social demand for a particular kind of virtue, self-sacrifice. Conventionally implied in the feminine "virtue" of self-sacrifice was a transaction, that is, giving up their own desires in exchange for their social "place."[69] However, since both took up the highly-regarded social status of physician, one wonders where their virtue of self-sacrifice would lead them? Within the medical profession, gynecology and obstetrics were conventionally a low-esteem specialty, and women medical students were often assigned to these departments by their male professors. Yang was a surgeon and had proved to be as good as other general medical practitioners before she entered the disciplines of gynecology and obstetrics. Her sacrifice of such a prestigious career for a calling to save more women won her a sort of heroic recognition from her male colleagues for her courage to take up such a challenging task. Meanwhile, different from the reception of Yang's feminine heroism and courageousness, Yoshioka was constantly battling with gender discrimination from her colleagues and even patients. The virtue of Yoshioka's self-sacrifice was to exchange for the recognition of professional authority that I am going to explain next.

Like Yang, Yoshioka was born into a physician's family. Yoshioka's father was a doctor of traditional Chinese medicine and mastered Chinese classical studies. During the transition from the Bakumatsu to the Meiji era, he was one of the few traditional doctors who combined eclectically the new styled stethoscope and other examination instruments for assuring accurate diagnosis. From her childhood Yoshioka served as an assistant in the family clinic, preparing prescriptions and herbal drugs. She recalled a couple of episodes in her initial feminist enlightenment of women's rights in her primary-school era. As soon as she knew her younger sister was bullied by a chap from the same school, she ran to beat up the boy, who ended up quitting the school. In another episode, she protested against teachers who patronized female students and granted only male pupils the duty of doing routine clean-up. Moreover, in her composition class, she wrote a critique charging that the violence of male historical writing incurring to a noblewoman false charges of reprehensible immorality, was even worse than the physical bullying of women.[70]

If Marian Yang formed her professional identity through the clinical experiences of women patients and the recognition of herself as a woman, Yoshioka established her own identity from historical inspiration, through a genealogy of Japanese women physicians. In her own account of the history of Japanese women physicians, Yoshioka documented the interconnection between female physicians and the realm of female bodies perceived as an object of knowledge. The earliest reference to women physicians was found in the legendary stories of the *Kojiki* (Annals of Ancient Events, or Records of Ancient Matters), and the Japanese official definition of women physicians (*joi*) in the legal commentary *Ryô no gige*[71] in the eighth century specifies,

> Joi were servants of the government who were selected as candidates from the age group between 15 and 25. Thirty intelligent and qualified women were selected every 6 years. These candidates were taught by oral instruction the subjects of miscarriage prevention, difficult labor, bone fracture, external injury and acupuncture, and were given examinations monthly by the physician officers and annually by the pharmacy bureau. Their training was completed in seven years.[72]

Since the specialties of women physicians were defined officially as within the aforementioned subjects, their social and professional status was not as highly esteemed as that of their male counterparts. More often than not they were regarded as not much different from midwives, acupuncture specialists, or nurses. Since they received mere oral instruction without reading directly the texts (they might be illiterate when they were selected for training), what they could improve most proficiently was limited to the level of technique, instead of medical theory. This asymmetric distribution of knowledge, Yoshioka commented, had to do with the power structure of gender inequality. Unlike her contemporary feminists, Yoshioka did not express militant antagonism toward the historical sedimentation of male-centeredness. To some extent, she wanted to bring to the readers' attention that it was through the hand of the State that the institution for female-physician training was able to come into shape.[73]

According to Yoshioka, the above official history was not able to resurface in the documentation until the unification of Togugawa Ieyasu (1542–1616), who ended an extensive period of warring states in 1603. Thanks to Ieyasu's mandate of a thorough search for ancient documents and their recompilation, the legal commentary could be partially recovered, tracked by later medical scholars, and finally published in the Bakumatsu era. This legal document thus became the basis of authority for the Meiji government to certify the profession of *joi*. Since the documents were well preserved after the Togugawa period, the profiles of these women physicians were much clearer.

The first *joi* in Yoshioka's account was Nonaka Enko (1661–1728), who was famous for her legendary talent of pulse reading through a fine thread. Enko was the daughter of Nonaka Kenzan (1615–1663), a Confucian scholar and the chief retainer in the Tosa domain. In 1664, their enemy took advantage of Kenzan's death to confiscate the Nonaka possessions and force Enko and her brothers into exile elsewhere. In her life of exile in Sukumo, Enko learned to read Chinese classics and poetry to relieve the pain from such traumatic disasters. She was released from an almost forty-year of confinement at the age of 44. When she was allowed to return to her hometown Kôchi, she established herself as a lauded physician. She would examine and treat poor folks and common townsmen in person. However, she would refuse to see samurai in person and demand their wrists be tied with a thin thread in order to read their pulse for diagnosis. This unusual antagonism against the samurai, Yoshioka speculated, stemmed from the painful mistreatment Enko had experienced in the Tosa domain. Enko, even though penniless, refused to receive any financial support from the enemy or accept marriage propositions from other domain leaders, despite the advice from her loyal ex-servants. She would not compromise for the sake of mere physical comfort and forsake the family glory. She remained single all her life, and established a family shrine and monument in order to recount her ancestors' integrity and the servants' loyalty in the domain history. In an anecdote describing her eccentric character, she often dressed like a woman knight-errant, carrying a sword, and wearing a veil without make-up, when she went out for a horse-ride. Her indisputable integrity was acclaimed to have magical efficacy in curing eye diseases.[74] From this narrative, we find a more fruitful picture of the profession of women physicians. It goes beyond the legal definition from the *Ryô no gige* that women physicians specialized merely in obstetrics, acupuncture, external injuries, and treatment of fractures.

Another renowned woman physician active around the same time was Watarai En'nyo (1664–1726), who was highly regarded not only for her medicinal techniques but also for her chaste virtue. Like Enko, she also specialized in ophthalmology. She became a Buddhist nun in her old age. Again through En'nyo, Yoshioka was presenting another impeccable model for women physicians, a model showing that knowledge and morality form an integrated whole. Morisaki Hoyû is another name worth mentioning because of her contribution to the production of women physicians in obstetrics and her production of a medical text in obstetrics. According to Yoshioka, En'nyo and Hoyû serve as two examples of how women received their training, one from family ties such as father and husband, and the other from a master-disciple relationship. For Yoshioka, these two female doctors were exalted beyond the status of mere technicians and gave a moral model for emulation.[75]

However, the term *joi*, as it appeared in the fiction of the popular novelist Ihara Saikaku (1642–1693), was different. Ihara, who often wrote realistic stories and strange folktales with valuable sidelights on the economic life of Edo-period townsmen and samurai, associated *joi* with the profession of abortion, called *Chûjôryû* under the disguise of menstruation treatment.[76] As judged by Yoshioka, the abortion association stigmatized the *joi* profession and a history once glorified by Nonaka Enko, Watarai En'nyo and Morisaki Hoyû, only because some of the female physicians abused the use of reproduction knowledge and helped spread a lustful quest for sexual pleasure in the Edo period. This perceived dark side of *joi* history, although contributing to the temporary alleviation of overpopulation pressure in the Edo period, developed an infuriating moral corruption in the society and became a mischievous feudal past that Yoshioka was rejecting in favor of the Meiji era of Enlightenment.

If Marion Yang attempted to get closer to midwifery, Yoshioka tried to get away as fast as possible from the stigma of midwifery, which was equivalent to abortion. For her, in order to develop a new history of *joi*, it was important to disassociate from this blemished past and locate another source for historical inspiration. Yoshioka found the genealogical line of Dutch Studies. For instance, the wife of the great master Ogata Kôan (1810–1863) was the daughter of the Dutch Studies pioneer Unakami Zuiô (real name Inamura Sanpaku, 1759–1811), and she acquired the training of Dutch medicine from her own father. Two other female physicians were Kusumoto Ineko, a daughter of the Dutch physician Philip Franz von Siebold (1796–1866), and Hikita Chimasu, both known for their specialty in smallpox vaccination in addition to obstetrics. This positive and fruitful Dutch experience prepared for the unfolding of *joi* history in the Meiji period. Yoshioka found a contrast to the Dutch *joi* Ineko in the traditional Chinese medicinal *joi* Takaba Ranko, whose personality had a flavor similar to that of Nonaka Enko, full of masculine heroism, often masquerading in male clothing and devoted to politics and education.[77] These personality traits reemerged as a pattern throughout Yoshioka's genealogical accounts, and constituted a stereotypical feminist role model with which Yoshioka identified. If Marion Yang identified with the female body of frailty, physical suffering and motherly love, Yoshioka had a much more ambiguous identity: in addition to traditional chastity and family loyalty, she was longing for a transgressive possibility for masculine heroism, in the form of a cavalier knight-errant, to rescue women from all kinds of torment.

As to the Japanese *joi* history during the Meiji enlightenment, Yoshioka's genealogy assigned Ogino Ginko (1851–1913) as the pioneer *joi* who marked the watershed of a new type of women physician who first entered the State's bureaucratic system through a national examination for professional certificates. According to Yoshioka, Ogino's personal experience of becoming a physician made a good case for the modern need for female physicians. In 1875–79, Ogino was an excellent student at the Tokyo Women's Normal School, and was planning to be an educator. She befriended the then feminist activist Shimoda Utako, who later founded in 1899 the Jissen Girls' School and Women's Arts and Crafts School, from which many Chinese women received their overseas education, as I mentioned in the introduction. In addition to encouragement from Shimoda, the force driving Ogino to shift her professional interests towards medicine was the humiliating experience of contracting venereal disease from her husband and being treated by male physicians for as long as two years.[78] One can imagine that venereal diseases were often socially stigmatized and associated with prostitution. Being treated by male physicians under such a depressing situation was even more devastating for Ogino. We should think in her shoes that traditional Chinese treatment for such a disease would not go beyond pulse reading and observation of facial color. However, in the wake of the modernization of medicine, Western-style treatment would involve frequent physical check-ups of private parts. This would induce an intense sense of violation of dignity and body every time Ogino had to undergo them in her treatment process. Consequently, she felt "acutely" (*tsûkan*) the necessity of female physicians, and launched a brand-new career pursuit. Ogino was determined to save women from the same depressing illness so that they did not have to go through similar humiliating experiences.[79]

With the help and recommendation of Ishiguro Tadanori (1845–1941), a baron and the president of the Red Cross Society, Ogino was able to study Western medicine in Kôju Hospital from 1879. Due to her family's opposition and refusal to pay any school expenses, she had to make her own living. After graduating in 1882, Ogino encountered tremendous difficulties from the government under the pretext that there was no historical precedent for women to participate in the National Physician Licensing Examination. There was no way to make an exception for her. After two years of painstaking endeavors without any positive result, Ogino came up with two strategies to combat such an institutional impasse. First, she tracked down historical materials and unearthed the aforementioned 8th century legal commentary of *Ryô no gige*, to challenge the bureaucratic pretext of there being no historical precedent for her case. Second, she considered detouring for an overseas physician training and certificate, making a forceful comeback later for official recognition. Holding the historical evidence to back up her proposition, she visited the then director of the Hygiene Bureau, Nagayo Sensai (1838–1902, a student of Ogata Kôan), in order to negotiate for a chance in the future. Moved by Ogino's earnest determination and convinced by her historical evidence, Nagayo finally gave his permission. In 1884 he set up an unprecedented case qualifying women to participate in the National Physician Licensing Examination, together with thirty thousand Western-styled and traditional Chinese-styled physicians. In that year, among the three women candidates in the Examination, Ogino was the only one to pass. Ogino's institutional victory marked a most glorious page of *joi* history.

Thanks to this very significant event, Yoshioka was able to figuratively give birth to the establishment of Tokyo Joshi Igaku Semmon Gakkô (Tokyo Women's Physician Professional School, a forerunner of today's Tokyo

Women Physicians University) sixteen years later in 1900. Ogino, and other female physicians also, had presented their profession as an indispensable medical practice in their petition for the National Physician Licensing Examination. The petition in 1883 addressed an acute clinical problem. Maternal care was crucially needed during pregnancy, and yet very often women were ashamed to report any symptoms or discomforts of their private parts to their husbands in order to seek medical help, because they felt disgraced to have such an intimate check-up inside their reproductive organs. However, without a thorough physical examination, there was no way for correct diagnoses and appropriate measures. This clinical problem imposed a grave difficulty for medical treatment. An initially minor and easily-treated gynecological illness often turned into a severe one without treatment. The petition pleaded the need for a women physician service to treat their fellow women's gynecological illnesses.[80] This clinical problem was so prevalent that it went beyond national boundaries. Yoshioka noted that she often encountered and treated Chinese female patients in her clinic for this particular reason. She also welcomed Chinese women from China, Manchuria and Taiwan to study in her medical school in the decades between the first and second Sino-Japanese Wars to foster international friendship between China and Japan.[81]

History played an ironic role. On the one hand it helped justify and support the argument for the institutional need for female physicians. Yet on the other hand it defined and marked confining boundaries, in that women physicians were unable to strip off the labels of gynecologists and midwives. Yoshioka lamented that when she opened up her own clinic Shisei Iin in 1897, there were only a few patients, since her name was not well known and the public perception of *joi* made no distinction from nurses and midwives. In order to change this gender discrimination, she adopted the tactics as her male colleagues: charging the same fees, dressing in trousers and serving on call every night.[82]

In addition to the aforementioned gender bias, an incident in 1908 devastated the hard-fought gender battle. The year 1908 was the ninth year after Yoshioka's Women's Medical College was founded in 1900 and was able to produce its first graduate, Takeuchi Shigeyo. Yoshioka planned to use the occasion of the first convocation ceremony to propagate the importance of women physicians and gain public recognition. Among the guest speakers invited from various universities, the most glamorous was the marquis Ôkuma Shigenobu (1838–1922). Before he inaugurated the speeches of congratulations, a group of male journalists, who shared interest with the conservative clique of physicians, took over the podium and roared out with anger: "Women who receive higher education are either married late or remain single throughout their lives. They are reducing the number of the Japanese population. They are going to bring about national crises." Other chauvinist howls followed: "Women physicians cutting open people's bodies with bleeding surgical knives totally damages traditional Japanese good morals and manners (*jûnpû bizoku*)!"; "Their polluted menstrual blood violates the sacredness of the operating room!"; "They are not reliable to have human life entrusted into their hands, since they need breaks during periods of pregnancy!" They concluded that women physicians would ruin Japan (*joi bôkoku*). Ôkuma finally shut off these appalling outbursts, when he stepped on the podium and said forcefully, "Gentlemen, today's vehement debates on women's higher education and the future of women physicians could carry on until tomorrow morning. But you would still be unable to come up with a solution momentarily or any time soon. Why not give yourselves ten or fifteen years to observe the facts of their development as to the result of whether women physicians are adequate or not." Ôkuma's statement broke the deadlock of arguments at the convention, and the crowd finally dispersed quietly.[83]

Hence, the whole profession of women physicians became subject to the crucial experiment of "time." In other words, male conservatives were watching whether the Japanese *joi* would misstep along the way, which would give them the chance to eliminate the profession. A decade or so later, when Margaret Sanger brought in contraceptive technology, Yoshioka as the director of the Japanese Women Physician Association (*Nihon joi kai*) was the only woman representative strenuously opposing the birth-control campaigns. According to Yoshioka's memoir, she invited only Japanese women physicians to her own hospital for Sanger's presentation and Ishimoto Shizue's interpretation. During the discussion, Yoshioka and other women physicians posed a number of questions for which Sanger was not able to provide an answer or defend her own position. The nature of the actual questions was not recorded. The embarrassment caused by antagonistic questions hurt Ishimoto's feelings and engendered her resentment of Yoshioka's disagreeable and uncompromising position. After this event, the women's journal *Fujin sekai* carried a caricature entitled "Ten-Years-later Yoshioka Yayoi," in which Margaret Sanger was standing up while Yoshioka was kowtowing at her feet. The caricature's text ridiculed Yoshioka as an idiot because she was rejecting the money-making opportunity of birth-control counseling. Ironically, as Yoshioka pointed out in her postwar memoir, these Japanese women activists in the birth-control campaign became the most fervent propagandists of the Kôseishô (the Ministry of Health and Welfare) campaigns of "More Births, More Children, More Soldiers!" during the wartime period.[84]

There were two main reasons for Yoshioka's opposition to birth control. First, she saw a great danger for chastity. She argued that childbirth was not a mere family matter, but a greater concern for national interest and strength. Furthermore, birth control had tremendous negative impact on public hygiene and social customs, since contraceptive methods provided freedom for sexual gratification, and would eventually lead to a hedonistic mentality and corrupt the original good intention of chaste principles and the production of better progeny. Second, birth control would inevitably be diverted from its primary purpose of saving the poor, as advocated by Margaret Sanger, because the poor would lose any motivation to work for their children, and subsequently acquire indolence and ill habits such as drinking and idleness. Yoshioka's anticipation of such devastating consequences precluded her from endorsing the birth-control campaigns. She agreed only that, under circumstances such as inheritable diseases, tuberculosis and ill health of mothers, physicians should prescribe birth-control methods.[85]

In addition to the two explicit reasons from Yoshioka's argument, I would like to explore other aspects of why she was against birth control. As a director of the Japanese Women Physician Association and the chancellor of Tokyo Women's Medical College, she was in a struggle for official, public and professional recognition. Even though women physicians could make a fortune from birth-control services, money could not buy respect from their male colleagues, especially those conservative male physicians who for a decade had been waiting to find fault with women physicians in order to stir another uproar. As I demonstrated above, Yoshioka was also conversant with the historical development of *joi* and was aware of its previous stigma of practicing abortion and fostering a lust-driven culture in the Edo period. It was easier for feminists to embrace birth control without reservations, but it was harder for female physicians since they were the persons who would encounter the clinical cases in their daily medical practice.

On the issue of sexuality and morality, Yoshioka's personal experience, of her mother's suffering from her father's infidelity and extramarital affair with a money-thirsty young maid next to their family clinic, speaks strongly for her position on chastity. This unfortunate incident not only ruined her parents' love life and family harmony, but also bankrupted her father's credibility, accumulated from his lifetime of medical practice as a respectable physician in a communal village. Watching her mother's tearful face and her father's career failure, Yoshioka's compassion was not merely expressed for her own family tragedy, but also for other families in Japan.[86] Hence, her position against birth-control campaigns was acquired from concerns for professional security and personal compassion as well. Whereas her compassion for family tragedy and women's suffering moved her in the anti-birth-control direction, Marion Yang's compassion for women's suffering harmonized with her own feminine identity and led to overwhelming support for birth-control work.

CONCLUDING REMARKS

When I first watched the Chinese movie *Raise the Red Lantern* years ago, my impression was that it revealed a typical indifferent male perspective on women's struggle and frustration within the physical confinement of a household, and the mental bondage of a patriarchal morality such that the mistresses from different class backgrounds turned against one another. However, as I watch it now, the scheming battles among the mistresses to win the master's favor actually illuminate four options of the conventional survival strategies of women confined in the household. These options include: the virtuous transaction of self sacrifice and domestic tolerance, the transaction of service, the transaction of theatrical talents, and the transaction of beauty and education, embodied respectively in the four mistresses in that order in the movie. All of these survival strategies are aimed finally at the biological function of reproducing male progeny as a sexual exchange for their household security and the patriarch's approbation. In comparison with these four survival strategies, both Yoshioka's and Marion Yang's virtue of self-sacrifice worked as a transaction for professional security, acclaim and credibility, and furthermore for their professional identity and self-identity. Nonetheless, their positioning on morality was pushed so far that the issue of sex was completely out of the picture. Ironically, this asexual morality of self-sacrifice helped them deter the eugenic mission of biological reproduction, which held that women like Yang and Yoshiioka should reproduce as much as possible.

As true believers of the national pedagogy, Yoshioka and Yang were not different from other women who were monitored by the patriarchal moral codes. However, their pursuit of higher and professional knowledge provided a leverage to transcend household confinement and its final destination of reproducing biologically the patriarchy and nation, and subsequently severed the correspondence between the female body and the national body. Moreover, as they acquired the reproductivity of knowledge, they in turn possessed an interpretive authority on morality. In the biographical piece on Yang's life, her total devotion of every inch of physical and mental strength to the extended female body became a moral aspiration for other followers on the same path. Yang's marriage to and romanticization of her gynecological career in maternal health and childcare, and reproduction of generations of midwifery helpers and specialists as her own offspring, served as a counterexample against the eugenicists' hard-line agenda.

In the same ironic fashion, Yoshioka played an active role in the discursive production of women's virtue (*futoku*). At the time of the Japanese Greater East Asian War (*Dai Tôa sensô*), Yoshioka commented that the conventional moral demand for women's great sacrifice as good wives and wise mothers required unrealistic blind devotion and produced unnecessary waste. She defined the real spirit of sacrifice (*gisei*) as the true perception and resolution (*kakugo*) of finding a right path appropriate to one's virtue. She said, "One has to think thoroughly about what one is good at and how to make the best out of it. Once one realizes the right path, work selflessly with might and main. To serve by making use of every wisdom is what I consider the right path befitting women's virtue."[87] Accordingly, the virtue of housewives was to function properly in the roles of good wife and wise mother within the household. The virtue of career women was to make all-out efforts for professional achievement. Considering herself as a professional physician, Yoshioka devoted her effort to taking care of patients and helping them go back to productive work as soon as possible. More importantly, she was helping those housewives to reproduce plenty of healthy and adorable children to meet the wartime need.[88] Yoshioka detached herself from the biological function of progeny reproduction, and redefined women's virtue in terms of individual capability and its different function according to their status in society. Hence, Yoshioka defined not only her own performative role but also that of other female physicians as her followers, and rendered eugenic pedagogy into empty words, since elite women like Yoshioka and other physicians were the best eugenic surrogate to reproduce superior national offspring.

NOTES

[1] Ono Kazuko, *Chinese Women in a Century of Revolution, 1850–1950* (Stanford: Stanford University Press, 1989), pp. 26–27.
[2] Ibid., pp. 54–55. Shimoda's Girls' Practical School was established in 1899 and began to admit Chinese women in 1901. By 1914 over 200 Chinese women students had graduated from the school.
[3] Ono Kazuko, "Shimoda Utako to Hattori Unokichi" in *Kindai Nihon to Chûgoku*, ed. Takeuchi Yoshimi and Hashikawa Bunzô, (Tokyo: Asahi shimbunsha, 1974), vol. I, pp. 201–221.
[4] See Elisabeth Croll, *Feminism and Socialism in China* (London: Routledge and Kegan Paul, 1978), for the frustration of Chinese women's sexual struggles when competing with the national struggles.
[5] Sharon H. Nolte and Sally A. Hastings, "The Meiji State's Policy Toward Women, 1890–1910," in *Recreating Japanese Women, 1600–1945*, ed. Gail L. Bernstein (Berkeley: University of California Press, 1991), pp. 151–174.
[6] Kang argued in his *Datong shu* (The book of great harmony) that because childbirth is an act of public significance in which a mother, as heaven's surrogate, gives birth to a human being, the surroundings naturally have to be prepared in full by society. See *Chinese Women in a Century of Revolution, 1850–1950*, pp. 40–46.
[7] See Ru-chiang Su, "Birth Control in China" (Master's Thesis, University of Chicago, 1946), pp. 152–153. Other positive aspects include prevention of illegitimate birth, prostitution, over-population as social consequence of the individual need for social security, national prosperity, and world peace.
[8] Ibid., pp. 8–11.

[9] Margaret Sanger, *Women and the New Race* (New York: Brentano's, 1920), p. 94.
[10] Ibid., pp. 179–182.
[11] Ibid., p. 183. It is not clear what Sanger meant by "race."
[12] The most recent study is Dorothy Roberts' *Killing the Black Body: Race, Reproduction, and the Meaning of Liberty* (New York: Pantheon Books, 1997), which argues that as the birth-control movement veered from its radical feminist origins toward a eugenic agenda, birth control became a tool to regulate the poor, immigrants and Black Americans. See the chapter "The Dark Side of Birth Control," pp. 56–103. For a discussion of Sanger's eugenic rhetoric, see Marouf Arif Hasian, Jr., *The Rhetoric of Eugenics in Anglo-American Thought* (Athens: the University of Georgia Press, 1996), pp. 86–87.
[13] Margaret Sanger, *My Fight for Birth Control*, reprint of the 1931 edition, (Elmsford: Maxwell Reprint, 1969), pp. 271 and 238–254.
[14] Ibid., pp. 255–270.
[15] See "Ying Sangeer furen" (Welcome Mrs. Sanger), *Funü zazhi*, vol. 8, no. 6, pp. 2–3.
[16] Se Lu "Chaner zhixian yu zhongguo" (Birth control and China), *Funü zazhi*, vol. 8, no. 6, pp. 10–14.
[17] Chen Dezheng "Hunyin he shengyu" (Marriage and childbirth), *Funü zazhi*, vol. 8, no. 6, pp. 84–89.
[18] These three translated articles can be found in the same issue, pp. 37–48.
[19] Ôta Tenrei, *Nihon sanji chôsetsu hyakunen shi* (The hundred-year history birth control in Japan), (Tokyo: Yamahan kagaku, 1976), pp. 176–181. According to Katô Shizue's autobiographical accounts, since the birth-control devices were under the cover-up of STD vaccines and without any specific instructions, some people did not know how to use them. There were quite a few instances in which patients mistook the "Seimori" suppository as oral medication and did not receive a desirable effect. See her autobiography, *Aru josei seijika no hansei* (Tokyo: PHP Kenkyûjo, 1981), p. 152.
[20] Ibid., pp. 145–6.
[21] Yamakawa Kikue, "Kyôsan Shugi no noroi" in *Sengo boseino yukue*, ed. by Suzuki Naoko (Tokyo: Domesu Shuppan, 1985), pp. 38–44, and "Shin Marusau shugi zehi" in *Shûkan fujo shinbun*, 14 April, 1921, pp. 5–7.
[22] According to Sumiko Otsubo, Hiratsuka was inspired by the Swedish feminist Ellen Key (1849–1926) for her ideas of motherhood protection and eugenics. See her "Eugenics in Imperial Japan: Some Ironies of Modernity, 1883–1945," (Dissertation, The Ohio State University, 1998), chapter IV, "Hiratsuka Raichô: Japanese feminists and marriage restriction legislation," pp. 143–201.
[23] Hiratsuka Raichô, "Sanji seigen no mondai" in *Sengo boseino yukue*, pp. 33–37.
[24] Yosano Akiko, "Saikin no zakkan yori" in *Sengo boseino yukue*, pp. 44–46.
[25] Ishimoto Shizue, "Sanji seigen wa naze hitsuyô ka" in *Sengo boseino yukue*, p47–51.
[26] Pan's article "Shengyu xianzhi yu yousheng xue (Birth control and eugenics)" in June, 1925 in the *Lady's Journal* was included in Pan's collected works *Pan Guangdan wenji*, vol. 1, pp.349–359.
[27] Naigai Hisomu, "Sanji seigen wa minzoku jisatsu" in *Dô to shizen* (Tokyo: Jinbun shoin, 1936), pp. 106–108.

[28] Zhen Fu, "Lu Xun he Zhang Xichen" in *Dushu zazhi*, vol.1 (1979), no. 1, pp. 87–88.
[29] Zhang Xichen, "Xin xing daode shi shenmo?" in *Xin xingdaode taolunji* (Shanghai: Kaiming, 1925), pp. 16–18.
[30] Ibid., pp. 24–40.
[31] Ibid., p. 98.
[32] Ibid., pp. 168–176.
[33] See *Xin wen hua*, vol. 1, no.1 (January, 1927), pp. 116–117.
[34] For a much detailed biography of Zhang, see Charles L. Leary, "Sexual Modernism in China: Zhang Jingsheng and 1920s Urban Culture,"(Dissertation, Cornell University, 1994).
[35] Cit.note. 31, pp. 128–129.
[36] Ibid., pp. 33–41.
[37] Ibid., pp. 106–107.
[38] *Xin wen hua*, vol. 1, no.2, pp. 19–37.
[39] Ibid., pp. 19–37.
[40] Ibid., pp. 127–130.
[41]Pan Guangdan, "'Xin Wenhua' yu jiakexue: bo Zhang Jingsheng" in *Pan Guangdan wenji*, vol. 1, pp. 401–406.
[42] Ibid. Also August Forel, *The Sexual Question: A Scientific, Psychological, Hygienic, and Sociological Study* (New York: Medical Arts Agency, ca. 1914), pp. 58–59.
[43] *Pan Guangdan wenji*, vol. 1, pp. 406–408. These two articles were originally published in the newspaper *Shishi xingbao*, on May 5 and June 24, 1927.
[44] This article was collected as an appendix to Pan's Chinese translation of Ellis' *Pyschology of Sex* (Xing xinlixue) (Beijing: Shangwu, 1997), pp. 616–715.
[45] Ibid., pp. 624–625.
[46] Ibid., p. 620.
[47] See Pan's *Zhongguo zhi jiating wenti* (Chinese family problem), in Pan's *Collection*, vol. I, pp. 153–154.
[48] Naigai Hisomu, "'Fujin' toiu idai naru te" in *Dô to shizen*, pp. 222–227.
[49] See *The Chinese Medical Journal*, vol. 48 (1934), no. 7, pp. 786–791.
[50] See Su's *Birth Control in China*, Chapter 5, especially p. 215 and the concluding chapter.
[51] This series of documentaries were produced by China International TV Corporation in 1999. The statement was provided by Zuo Qi, one of the first generation midwives trained by Marian Yang and I was unable to verify her words through the PUMC files at the Rockefeller Archive Center.
[52] See Tamura Toshio, "Sangâ fujin no sanji seigen jikkô hôhô o yomite," originally published in *Osaka Asahi shimbun* on April 10, 1922, and later included in *Yamamoto Senji zenshû* (Tokyo: Chobunsha, 1979), vol. 3, pp. 22–24, attached to Yamamoto's critique of Margaret Sanger's contraceptive methods.
[53] *Sangâ joshi kazoku seigenhô hihan*, which was banned after it was published, and was included in *Yamamoto Senji zenshû*, vol. 3, especially, pp. 20–21, Yamamoto's comment on Tamura's reading on Sanger's contraceptive instruction.
[54] Ronald P. Loftus, "Female Self-Writing: Takamure Itsue's *Hi no Kuni no Onna no Nikki*" in *Monumenta Nipponica*, vol. 51 (1996), no. 2, pp. 153–170.

[55] See the membership list of the Racial Hygiene Association (*Minzoku eisei kyôkai*) in *Minzoku eisei*, vol. 4 (1934), no. 1, pp. 87–101.
[56] Also see Sally A. Hastings' biographic article, "Yoshioka Yayoi" in *Doctors, Nurses and Medical Practitioners—A Bio-Bibliographical Sourcebook*, ed. by Lois N. Magner, (Westport: Greenwood Press, 1997), pp. 315–319.
[57] See *Zhongguo Xiandai kexuejia zhuanji* (Biography of modern Chinese Scientists) (Beijing: Kexue chuban, 1991), vol. III, pp. 552–559.
[58] This biography was edited by Yan Renying under the title *Yang Chongrui Boshi* (Doctor Yang Chongrui) (Beijing: Zhongguo Xiehe yike daxue, 1990), as a memorial piece for her hundredth birthday, and was contributed to by her old colleagues, students, and family who promoted birth-control campaigns in the 80's.
[59] Yang Chongrui (Marion), "Wo de zizhuan" in *Yang Chongrui Boshi*, pp. 143–153.
[60] *Yang Chongrui Boshi*, p. 155.
[61] See Xiong's article "Zhongguo jinshi de xinshenger zhaohu" in *Zhongguo jinshi shehui wenhuashi lunwenji*, (Taipei: Academica Sinica, 1992).
[62] Yang, "Chanke jiaoyu jihua (The plan for midwifery education)," *Yang Chongrui Boshi*, pp. 136–142.
[63] Ka-che Yip, *Health and National Reconstruction in Nationalist China: The Development of Modern Health Service, 1928–1937* (Ann Arbor: AAS, 1995), p.59.
[64] *Yang Chongrui Boshi*, pp. 16–17.
[65] Ibid., pp. 125–136. The English version of Yang's report can be found in "Birth Control in Peiping," *Chinese Medical Journal*, Vol. 48, no. 8, (August, 1934), pp. 786–791.
[66] *Yang Chongrui Boshi*, pp. 34–35.
[67] Lei Zhifang, "Woguo jihua shengyu de tuohuang zhe," in *Yang Chongrui Boshi*, pp. 15–19.
[68] Yang Guang, "Ta de xinfei zhi kaixiang yimian," in *Yang Chongrui Boshi*, pp. 110–114.
[69] Rey Chow, "Virtuous Transactions: A Reading of Three Stories by Ling Shuhua" in *Modern Chinese Literature* (vol. 4, 1988), pp. 71–85.
[70] Ibid., pp. 49–51.
[71] The *Ryô no gige* was the official commentary for the legal statutes Taihô ritsuryô stipulated in 701 on various subjects of bureaucracy, land and tax, and penalty. For historical details, see *Nihon joi shi* (The history of Japanese female physicians), (Tokyo: Chûo kôron, 1962).
[72] Yoshioka Yayoi, *Yoshioka Yayoi den* (Tokyo: Nihon Tosho Center, 1998), p. 87. This is a reprint from its first edition of 1941.
[73] Ibid., pp. 87–88.
[74] Ibid., pp. 91–94.
[75] Ibid., pp. 95–98.
[76] Ibid., 98–102. *Chûjôryû* was a specialized name for abortion techniques and named after Chûjô Tatewaki ,who was the forerunner of this field in gynecology and obstetrics in the period of Toyotomi Hideyoshi (1536–1598).
[77] Ibid., pp. 102–107.
[78] Ibid., pp. 108–109.

[79] Ibid., pp. 109–110.
[80] This petition was written by Ikusawa Kuno, who was the second *joi* to receive the physician certificate after Ogino Ginko. See *Nihon joi shi*, pp. 81–82.
[81] See Yoshioka's *Kono jû nen kan* (At the time of this decade) (Tokyo: Gakufû shoin, 1952), pp. 22–23.
[82] *Yoshioka Yayoi den*, pp. 223–224.
[83] Ibid., pp. 268–274.
[84] Yoshioka, *Kono jû nen kan*, p. 83.
[85] Yoshioka, "Genzai no nihon ni wa dai kinmotsu" in *Fujin sekai*, vol: 17, no. 4, (April, 1922), pp. 103–104.
[86] *Yoshioka Yayoi den*, pp. 192–193.
[87] Yoshioka, *Fujin ni atau*, (Tokyo: Ôzorasha, 1997, reprint, originally published by Hôdô shuppansha, 1943), pp. 4–5.
[88] Ibid., p. 11. Another irony which was diametrically opposed to the male eugenicists' hard-line agenda was that Yoshioka married her German-language teacher Yoshioka Kôta (1868–1922), who died of diabetes, a common hereditary disease already known at the time. When Yoshioka decided to marry him, none of the health considerations came across her mind. According to her own account, the elements she considered for marriage were: 1) she did not want to stay single; 2) she wanted to go to Germany to pursue and advance her medical profession and Yoshioka as a German-language teacher could help her with it; 3) Yoshioka's quiet and calm personality could complement her restlessness. Their marriage yielded a result of only one son. See *Yoshioka Yayoi den*, pp. 209–210.

5

EUGENICS IN THE SECOND SINO-JAPANESE WAR AND POPULATION POLICIES

WARFARE, ACCORDING TO HERBERT SPENCER, HAD VARIOUS BENEFICIAL EFFECTS to raise human organizations to a higher stage. First, it had the effect of extirpating races which were least fit to cope with the conditions of their existence, and exterminating relatively feeble tribes, tribes relatively wanting in endurance, or courage, or sagacity, or the power of cooperation. Warfare thus tended to maintain the life-preserving powers possessed by men. Second, the average advance caused by destruction of the least-developed races and the least-developed individuals was inheritable, thus providing a base for further development. Third, it stimulated technological development in responding to the imperative demands of war. Fourth, it brought forward the formation of large societies and brought about a social aggregation which furthered the development of industrialization, and during this stage of development war was no longer advantageous.[1] The Spencerian presentation of warfare as group selection contradicted the usual theory of eugenicists who saw warfare as a matter of individual selection. For instance, in a major speech in 1935 Hitler, echoing the influential Nazi racial hygienist Alfred Ploetz, warned that "every war goes against the selection of the fittest." This is because the best stock of men of large physique and strong constitution tended to be put on the front lines, while the worst ones tended to suffer least from war, going on to proliferate with their own progeny.[2]

History takes a different twist. The Spencerian evolutionary conception of warfare was embraced by both the Japanese and the Chinese in their confrontation of the second Sino-Japanese War (1937–45), in that both perceived the War as a progression (J. *shinka,* C. *jinhua*) from a war of individual heroes and armed forces to a war of total mobilization.[3] Scientists in general and eugenicists in particular participated willingly in designing social policies as part of the war strategy, advancing technology, and most importantly, enrolling as role models of morality in the Spencerian forms of endurance, courage, sagacity and cooperation which they saw as key standards for the survival of the fittest race.

The Japanese state and eugenicists justified their imperialist expansion into Manchuria and China proper as a natural outcome of population growth. The geneticist Komai Taku voiced the notion of Japanese racial superiority and defended his participation in Japanese campaigns surveying natural resources in Southeast Asia as an aid to local development and to the Japanese leadership in the region, in their resistance to Western encroachment. Another geneticist Tanaka Yoshimaro considered racial hygiene an appropriate method of war preparation and national participation. The renowned entomologist Matsumura Shônen (1872–1960) advocated the analogous connection between species and nations in their competition for survival, and affirmed the War as not only inevitable but also necessary for future progress in his plea for "National Awakening."[4]

The Chinese eugenicist Pan Guangdan accepted the second Sino-Japanese War as an opportunity to test the Chinese national stamina (*minzu yuanqi*) or vitality (*minzu huoli*) by the evaluation of two indices, the quantity and the quality of population. The test represented a threshold, which Pan applauded through the metaphor of a coming-of-age ceremony: a young adult grows into his manhood, a necessary part of the process involving the ritual of suffering hardship in war. Pan said that from the War, the Chinese had learned to know their merits and demerits, and had adapted them into a strategy of guerrilla warfare and war of attrition.[5]

Pan's interpretation of the War revealed an intriguing epistemological shift. Initially the scientists behaved as observers of nature. For instance, Spencer observed the human world indifferently, whereas later scientists acted as part of the nature they studied. Pan Guangdan and the Japanese eugenicists enthusiastically joined in the war effort and the test of the survival of the fittest. Here, we notice a split signification of the fittest and the survivors in the shift from value neutrality to enthusiasm, and thus a restructuring of the scientists' self-identity and national identity. This shift was not merely a unique event for China and Japan, but a transnational phenomenon. The global expansion of eugenics technology encountered different local interests as war became the outcome of conflicting interests in the larger setting of World War Two.

Because both Japanese and Chinese eugenicists perceived the Sino-Japanese War as a war of population, and justified the War as part of the process of evolution, this chapter lays out both the eugenicists' representation and their involvement in implementation of the State's policy of consolidation into a single aggregate unit of evolution in China and Japan. It centers on the interaction of the Sino-Japanese War and population policies, and their effects on the development of the eugenics movements in Japan and China. The following sections analyze respectively the Japanese and Chinese eugenicists' involvement in the population policies as war strategy, which were the embodiments of their conception of an ethnic nation and racial body. The last section concludes with the postwar reflections of two eugenicists, Nagai Hisomu and Pan Guangdan, on the outcome of the War.

JAPANESE EUGENICISTS' INVOLVEMENT AND POPULATION POLICY: CONSTRUCTING THE RACIAL BODY AND THE ETHNIC BODY

As an outcome of the war system mobilization, the Kôseishô, the Ministry of Health and Welfare, was established in 1938. Its main function was to serve the need for "healthy citizens, healthy soldiers," and to realize the eugenicists' blueprint of securing a source of healthy soldiers both in quality and quantity. Before the birth of the Kôseishô, in 1932, the government was dismayed by the fact that of 20,000 soldiers sent by the army to Manchuria, 500 carried tuberculosis and had to be dispatched back to Japan. Hence, the army was very enthusiastic about any preventive measures with respect to all sorts of communicable and transmittable diseases. In 1936, the medical bureau of the army submitted a proposal for establishing an Eiseishô (Ministry of Hygiene). It was based on the idea of its superintendent-general Koizumi Chikahiko to construct a centralized administrative hygienic organ to elevate the declining level of physical strength and health of citizens and soldiers.[6]

In 1937 the army submitted a revised proposal to establish a Hoken Shakaishô (Ministry of Public Health). The second proposal, years before the Beverisge plan was proposed in Britain, included the idea of making Japan a welfare country. The proposal was accepted on July 9, 1937, two days after the outbreak of the Sino-Japanese War, but was subsequently suspended in the Privy Council, since in this equivocal initial stage of the war, Japan was speculating that it could win in as little as three months. In the meantime, since some of the Privy Council members regarded the name *Shakai* (social) as inappropriate (a connotation of socialism) and the name *Hoken* (public health) easy to confuse with *hoken* (insurance), the committee decided to change the name of "Hoken Shakaishô" to "Kôseishô." The words *kôsei* from the *Book of Document* (the Confucian text *Shujing*), denoting "enrich the people's life," were thought to be rather appropriate to the purport of the proposed "Ministry." After it became clear that the war would be a long one, and in order to prepare for a thorough plan of war deployment, the Konoe Cabinet decided to speed up the establishment of the Kôseishô and reactivate the proposal as of January 11, 1938.[7]

Structurally within the Kôseishô, the eugenics division was moved from the Civil Affairs bureau, where it had been in the first proposal, to the Medical Bureau, in the second proposal. Finally, after the establishment of the Kôseishô, it was moved to the Prevention bureau, in order to coordinate with the divisions of Chronic Epidemic Prevention and Plague Prevention. This structural rearrangement could be seen as an outcome of reconciling the eugenic and hygienic approaches. Moreover, together with other functionary organs—like the bureau of Physical Strength in charge of sports, prenatal care and national parks for sport facilities, the Hygiene bureau for matters related to public hygiene, the Social bureau for problems of social welfare, and the Labor bureau for labor administration and industrial hygiene—the Kôseishô should be seen as a realization of the historical synthesis of managerial efficiency in the Lamarckism versus Mendelism debates among eugenicists, hygienists, socialists and sociologists, to incorporate both genetic and environmental measures for racial improvement.

The Kôseishô not only took up many of the objectives of the Race Hygiene Association, but also sponsored the draft of the National Eugenics Bill. The Kôseishô also worked closely with eugenicists who did research on the Taiwanese, Korean, and Manchurian populations. Japan's population then was 73 million and China's 470 million, and Japan's birth rate had dropped from 36.2% in 1920 to 26.7% in 1937 as a result of the birth-control movement. As such, the Kôseishô as part of the wartime system was responsible for meeting the qualitative and quantitative demand for drafted soldiers, in order to sustain Japan's leadership throughout East Asia. The Kôseishô adopted both qualitative and quantitative approaches in the design of population policy.

Qualitatively, the improvement of the Japanese race was based on the eugenics programs and supplemented by marriage counseling and coordination of the various organizations that would help in arranging marriages. Quantitatively, in order to increase the already-declining birth rate, the Kôseishô in 1941 tried to lower the average marriage age by three years over the course of the next decade, and set as a goal an average of five children per family with the encouragement of tax exemption, loans, rationing privileges, etc.[8] Overseas, the eugenicists had to juggle the management of human resources in the Japanese colonies, where we find that the eugenic qualitative approach was intriguingly absent. In what follows, I will focus on the eugenicists' involvement, in which we can detect a complementary twofold approach, namely, blood and culture—that is, taking the identity of Japan as a racial body (*Yamato minzoku*) and as an ethnic body (*Nihon minzoku*) in the making of the State's population policies.

While serving as the director of the Medical College of Peking University in North China, the eugenicist Nagai Hisomu was participating in eugenic activities and serving on the public-health investigation committee in the Kôseishô. His earnest call in October, 1937 for better quality and quantity of reproduction exhorted the younger generation to model themselves after the ancestors' spirit of sacrifice and follow the principle of racial hygiene in preparation for the war, which would determine the survival of the Japanese culture and race.[9] He lamented that some features of modern civilization, such as late marriage and venereal diseases, selected the non-fittest. He attributed these phenomena to the morbid influence of individualism, especially among elite men and women who selfishly indulged in their own temporary sexual gratification without any concern for the wellbeing of the family, society or nation.

Nagai continued to construct the identified objectives of the Japanese race for which every Japanese individual should sacrifice. Based on the criteria of racial hygiene, Nagai defined three qualifications of the superior race: 1) a large population; 2) excellent racial quality; 3) purified racial blood. Even though both China's 450 million and India's 350 million populations could qualify in the first criterion, they could not pass the second because of lower racial quality. Other populations, such as Russia's 230 million and the USA's 130 million, might qualify under the first two parameters, but they fell short on the third. As for the population of the Japanese race (*Nihon minzoku*), he estimated there were 70 million in the homeland, 22 million in Korea, 5.2 million in Taiwan, 1.5 million in North China and 0.3 million in Sakhalin. In addition to these populations, there were 33 million in Manchuria united wholeheartedly as the same body (*isshin dôtai*); therefore, Japan had enough people to meet the first criterion. Nagai's conflation of the Japanese empire with the Japanese "race"(*minzoku*) was wrapped in an ambiguous national identity. For the first parameter of quantity, the colonized could be counted as part of the race/nation. Yet for the second and third parameters of racial quality and pure blood, the possessive case of "we the Japanese race (*Wa ga nihon minzoku*)" did not denote any inclusive corporeal sense of the colonized subjects. Nagai wrote,

> For the second criterion, the gifted superior racial quality of we the Japanese race could be witnessed in the historical evidence of thriving cultural assimilation of the best elements from the East and West, and the mission accomplished in becoming the model for all human beings. The greatest potential of a race would act out sufficiently in time of war, and Japan has made it so far. This achievement has to do with the superior germplasm organized from the historical interbreeding within the Japanese race (Yamato minzoku) and self-contained geographically inside the archipelago empire. As for the third parameter of purified racial blood, it matters not only corporeally but also spiritually. The kin connection of blood signifies above all the spiritual union which is the foundation for the Japanese family system, and thus pyramidally the emperor at the top is the great patriarch of the nation as a grand lineage. This is the same goal as the German Nazi's racial purification in their recent racial-hygiene movement."[10]

In the definitions of the three parameters, the Japanese term "*minzoku*" used above by Nagai in a sense of quantity, denoting the "Japanese empire" (or, in the Chinese eugenicist Pan Guangdan's term, "ethnic nation") was also used to denote quality, as in the "Yamato" race. This double standard or flexible definition for a superior race was taken for granted, as if the "Yamato" race had not itself been formed through the process of military expansion and forced intermarriage between the conqueror and the conquered.[11] Moreover, this historical precedent of the "Yamato" race was not able to serve as a justification for making the present "Japanese empire" into the future "Yamato" race. We have seen this in Nagai's discursive statement. It can also be seen in the absence of eugenic campaigns in colonies such as Taiwan and Korea, whereas there was simultaneously a Japanese *kôminka* (imperialization) movement during the wartime period 1937–45 in these two colonies.

As Chou Wan-yao shows in her studies, the Japanese *kôminka* movement was a series of mobilization campaigns aimed at "making the Taiwanese and Koreans true Japanese." The *kôminka* movement included a national-language movement, a name-changing program, a military recruitment campaign, and reforms of native religions and social customs.[12] Among these campaigns of spiritual transformation, one finds that the most obvious approach—assimilation via interbreeding between the colonizer and the colonized—was missing. The Japanese Kôseishô-sponsored National Eugenics Bill did not include Taiwan and other colonies for implementation.

This intriguing phenomenon is critically examined by Oguma Eiji. He points out that whereas racial intermarriage between Japanese and Taiwanese/Koreans was promoted as an immediately effective policy of assimilation by the colonial governments to further legitimate their rule, it was strongly opposed by the eugenicists, Marxists and anti-assimilation theorists. As early as 1915, Kawakami Hajime (1870–1946), a Marxist economist, denounced the colonial governments' encouragement of racial mixture because he feared that the purity of the Japanese race would be contaminated by intermarriage with inferior people like Taiwanese and Koreans. Kawakami's racist postulation was further elaborated by Tôgô Minoru, a high-ranking bureaucrat in the Government-General of Taiwan. He was a friend of another Marxist theorist, Yanaihara Tadao (1893–1961), who was favored by most 1930's Taiwanese intellectuals for his well-received book entitled *Teikokushugi-ka no Taiwan* (Taiwan under Imperialism) because of his advocacy of anti-assimilation. However, like Kawakami, both Tôgô and Yanaihara based their anti-assimilation position on the same fear of racial mixture and contamination of Japanese pure blood. Tôgô not only opposed the assimilation policy but also tried to push for racial segregation in Taiwan. He maintained that although Japan had been through a racial mixture two thousand years ago, it had attained ethnic purity and supremacy since then. He tried to warn his compatriots that Japan would step on the same path of decline as the Roman Empire by adopting the same assimilation policy and corrupting the nobility of the Japanese national character. His opposition created so much antagonism in the colonial government that he was forced to resign from his post.[13]

Nonetheless, Tôgô's anti-assimilation position was supported by the Kôseishô and later substantiated by the eugenicists who used sociological, ethnological and medical studies to prove the inferiority of the colonized in both physical and mental aspects. On the other hand, academic science was also used by the Government-General of Korea to support racial intermarriage. It presented evidence such as that Japanese and Koreans belonged to a group with the same origin (*Nissen dôso ron*) and hence the same unity (*Naisen ittai*),[14] and that the racial coefficient of blood types from serological studies confirmed that homeland Japanese and Koreans were racially approximate.[15] It is important to note that the *Nissen dôso ron* was also used to distinguish between Koreans and Taiwanese. Koreans, it said, should be more easily assimilated since Taiwanese were distant both racially and culturally from Japan. It was said to have taken forty years of time after annexation to achieve what had been accomplished in Korea in twenty years.[16]

Eugenicists, even from the same serological methodology, could derive the opposite conclusion, against racial mixture. The serologist and eugenicist Furuhata Tanemoto used his three anthropological principles of blood type[17] and his own study of the genetic frequencies of blood-type distribution in the Japanese archipelago. He traced a specific pattern of blood-type distribution prevailing through the homeland Japanese population and overseas Japanese in the colonies, although he found a slight variation among various localities in the homeland.[18] This distinct pattern was to be found nowhere else in the world except among Hungarians and Mongolians. Hence, Furuhata confirmed the general conviction that the Japanese race evolved from interbreeding among different races from the South and North, and thereafter regenerated a superior patriarchal race (*daikazoku minzoku*) centered around the emperor of unbroken lineage (*bansei ikkei*).[19] The eugenicist Nagai Hisomu's position of anti-assimilation was as clear: his argument that the superior race had to avoid interbreeding with other races, pointed as a historical precedent to the decline of the Roman Empire due to racial mixture.[20] In the same fashion, another eugenicist, Ikeda Shigenori, insisted that the eugenics movement had to protect the root of pure blood, i.e. the traditional family life and continuity of pure unbroken lineage, in order to maintain the supremacy of the Japanese race and the leadership of Japan in East Asia.[21] This particular position of safeguarding Japan's racial supremacy and leadership in East Asia, in terms of rhetoric and ambivalent identities (Japan in, out of and above Asia), was very much regulated by the discursive practice of pan-Asianism and the Japanese perspectives on Asia from dissociation to pan-Asiatic co-prosperity. Japan started these perspectives from the adoption of western civilization, whereby it became one of the world powers. Later it transformed itself into a cultural synthesizer of the best elements of both the eastern and western civilizations and thereby the model for the world and the leader in Asia against western invasion.[22]

As eugenicists and eugenics supporters were predominant in the staff of the Kôseishô, their opposition to racial mixture was taken up as the official position of the Kôseishô. But there was also another concern. On January 22, 1941, the Kôseishô promulgated, based on the decision of a Cabinet meeting, the "Outlines of Establishment of Population Policy." It responded to the fears, expressed by the press, that the colonized populations might not help the war effort wholeheartedly. Therefore Japan had to design a population policy for the home country that would improve the Japanese population in both quality and quantity in order to insure its national security and leadership in East Asia.[23] The Outlines abstracted the goals above and gave instructions for specific objectives, focusing mostly on the Japanese archipelago. Quantitatively, the plan targeted the increase of the homeland population to a total of 100 million in the year 1960; hence the Kôseishô encouraged reproduction of each couple up to an average of five births. It proposed reducing tax on parents, providing prenatal care, subsidizing medical and educational expenses, and for larger families, granting public awards and military ration. Abortion and birth control were to be absolutely prohibited. Other public hygiene and health plans were proposed in order to reduce the mortality rate. Qualitatively, the Kôseishô intended to spread evenly the population over urban and rural areas, propagate eugenic ideas, and impose physical and mental training on urban youths.[24]

The wartime activist eugenicist Furuya Yoshio joined the Kôseishô in 1939 in various positions in the bureaus of Physical Strength, Hygiene, Plague Prevention and Social Affairs and other investigation committees.[25] In addition to the legislation of the National Eugenics Bill (*Kokumin yûseihô*) and Physical Strength Improvement Bill (*Kokumin tairyokuhô*), Furuya participated enthusiastically in drafting population policy. Like Nagai Hisomu, Ikeda Shigenori and Furuhata Tanemoto, Furuya's position was persistently opposed to miscegenation He exhorted those colonial bureaucrats who promoted racial intermarriage to be more prudent in such a serious and dangerous matter. His considerations included his racial prejudice and fear of racial contamination, and his dialectical perspectives on Asia. He saw Japan moving from dissociation to pan-Asiatic co-prosperity. One of his concerns was "time"—the invincible biological Time that science could not overcome or reverse. For Furuya, the making of the "Yamato" race from racial interbreeding was not a repeatable historical process. The eugenicists were confident that it was impossible to compress two thousand years into the short span of one single generation—i.e. twenty years—putting all the desirable specimens into a controlled laboratory and hence producing the reproducible outcome needed so badly for the cause of war, the ultimate test for survival.[26]

Furuya further expounded his reasoning of the dangerousness of racial mixture in the colonies. For instance, in the case of China, the Han Chinese had a long history and deep-rooted strength of cultural assimilation, as seen in the sinicization of Khitan and other northern races when they conquered Han China. If Japanese mixed with the Han Chinese, Furuya was afraid, they would end up becoming Chinese. As such, it would be like wasting precious Japanese sperm in the Chinese blood, which flows like the vast Yangtze River. Furthermore, since Chinese endemic characteristics were so different from the Japanese, the immigration of Japanese peasants and the imposition of farming system in such an unfamiliar ecological environment was not necessarily feasible. More importantly, from the difficulty of sustaining a Japanese educational system for these overseas Japanese, Furuya foresaw the potential danger that they would sooner or later lose their identification with the metropolis.[27]

Furuya continued his discussion of "scientific evidence" of the racial inferiority of the hybrids, such as low reproductivity and a physical frailty susceptible to tuberculosis. Since there were counter-examples of hybrids being strong and healthy, Furuya reserved his judgment that the hybridity was necessarily and scientifically inferior. What he hypothesized was that hybridization from two disparate environments would cause incompatible combinations of germplasm and make it difficult for the hybrids to survive the challenges of a different environment. He presented his own case study of nearsightedness among the Japanese, and tried to account for the phenomenon of why many Japanese were nearsighted. He investigated the Japanese aborigines, the Ainu, for five years and studied other areas in the Archipelago to measure degrees of hybridity. He compared a sample of the Ainu of the far North, whose cranium shape was consistently similar, vis-à-vis the crania of Japanese who differed appreciably one from another, and found them racially pure and hybrid, respectively. Since most Japanese children came from parents with different cranium shapes, he thought that their optic axes would become asymmetric and difficult to adjust, and thereby they would easily develop nearsightedness. It was this incompatibility from hybridization that concerned Furuya most. For him, this was the utmost danger of racial mixture and should be taken seriously by any colonial assimilation policy.[28]

In his later work *Kokudo, jinkô, ketsueki* (National land, population and blood) of 1941, because he was partly speaking for himself and partly speaking from his position in the Kôseishô, Furuya was careful in choosing his rhetoric without showing a crude sense of racial prejudice. He evidently realized that a language of racial discrimination ran the risk of sparking anti-Japanese nationalistic sentiment in the colonies. Furuya had presented his own experience in the series of Kôseishô administrative meetings leading up to the legislation of the Eugenics Bill. The bill, in its originally proposed form, was named first "*Minzoku yûseihô*" (Racial Eugenics Bill) and only later corrected to "*Kokumin yûseihô*" (National Eugenics Bill). This change was precisely because of the fear that the term "*minzoku*" was so provocative, it would instigate unnecessary racial friction in the project of constructing the Greater East Asia Order and thereby jeopardize Japanese leadership. However, there was an overreaction to avoiding completely the use of "*minzoku.*" Subsequently, Furuya suggested toning down the exclusionist tone of the term "*minzoku*" while recognizing correctly its proper function in the nationalist movement for regional unity. He did this in his comment on the political mobilization speech of the Chinese collaborator Wang Jingwei (1883–1944), who related his rhetoric of nationalism to Sun Yat-sen's *Three Principles of People.*[29]

Furuya further elaborated his understanding of nationalism in the case of the German Nazis' nationalistic exclusion of Jews. He defended apologetically the Nazi exclusion act as being in fact what Germans called "*Rassenpflege* (racial cultivation)," an effort of self-purification of their own racial blood, and not an aggressive plan to extinguish racially the Jews. He took positively the Germans' battle against their own problematic predisposition, the spoiling Jewish blood. Furuya somewhat admired the Germans' resolution to enforce thoroughly the Sterilization Law even with the consequence of suffering a population insufficiency.[30]

As for the case in Japan, Furuya's view of the problem of nationalism rejected the idealistic conviction of non-differentiating pan-Asianism. That argument said that since the Japanese race was originally a hybrid, there was no reason to set up a racial blood boundary among the races in the Greater East Asia Co-prosperity Sphere, as long as they received direct imperial rule. Furuya argued that such a simplistic and non-discriminating egalitarian idealism ignored the real problems of different cultures and biological constitutions generated from the racial encounter in the project of building a Greater East Asia. Racial assimilation and intermarriage was the best example. Although rhetorically Furuya agreed that racial intermarriage was the best way to achieve assimilation in the colonies, he argued that between the Japanese and the other racial groups there were mathematically immeasurable differences of spiritual disposition and function, along with a disparity of cultural traditions and life styles. Since culture determined the variation of predisposition it would be difficult to overcome such an appalling cultural gap. People, like animals and plants, went through a process of natural selection when they lived in different climates and environments for a long time, and once they were transplanted and crossbreeding with others, they could no longer enjoy their previous prosperity. Furuya explicated that the Japanese race, even though it was a hybrid outcome of North and South ethnic tribes, was contained in the closed environment of the archipelago without unlimitedly interbreeding with outside races, and thereby took the shape of its present uniqueness. Furuya criticized those pro-assimilationists as being ignorant of selection theory and the concept of "Time."[31]

In the same rhetorical fashion, Furuya, although he did not take negatively the idea of *Naisen ittai* (Koreans and Japanese as the same unity), saw forced racial assimilation as an "unnatural" process that would in turn give rise to problems of racial friction and competition. He pointed out that the pro-assimilationists did not have any idea of what education should be provided for racially-mixed children. For Furuya, since the approach of *Naisen ittai* was not intended to repeat the same natural-selection process that the Yamato race had been through in the last 2000 years, it could never produce a similar result. As for intermarriage with the Chinese, Furuya believed that in a short time the Japanese colonizing population would be assimilated into the Han Chinese population. He concluded that nationalism at this historical juncture was necessary to inaugurate an internal movement for the formation of racial biology, which was perceived externally as a derivative development of exclusion as an aspect of foreign policy. In other words, Furuya's version of nationalism aimed to enforce the quality and quantity of human resources as an internal constituent of a race or, put another way, as a way to racialize nationalism and perceive the Japanese nation as a racial body. For him, dealing with quantity would be the objective of population policy, while quality was the eugenics plan. Overall, these two mutually supplementary approaches would help mass-reproduce the Japanese of superior quality, entrusted with the great mission of assuming leadership in Greater East Asia.[32]

Furuya and other eugenicists viewed racial identity in terms of pure blood. This eugenic position of pure blood as an internal cohesion for self-identity reached its consummation in 1942 in the confidential population proposal, designed by the Kôseishô staff at the demand of the Minister, entitled "Yamato minzoku o chûkaku to suru sekai seisaku no kentô" (the Study of the Yamato race as the nucleus of global policy).[33] This document recommended that the second generation of Japanese colonizers receive obligatorily a patriotic education in the metropolis. Based on the study of 1500 interracial couples of mostly Korean males and Japanese females, it emphasized negative characteristics of the hybrids, such as low intelligence, sexual compulsiveness, dependence, irresponsibility, feeble-mindedness, susceptibility to endemic diseases, slavish submissiveness and, most significantly, lack of patriotism. It warned that the racial mixture of assimilation would in turn destroy the unity of the Yamato race, downgrade Japanese culture and finally cause Japan to lose the consciousness and capability of being a leader. Hence, the Kôseishô persistently opposed the somewhat favorable approach by the State to Japanese-Korean intermarriage.[34]

In the meantime, the *Kôminka* movement in Taiwan and Korea mobilized and recruited during the wartime 724,443 Koreans[35] and 207,183 Taiwanese as forced laborers and soldiers.[36] Japan's need of Korea and Taiwan to provide human resources and military support on the one hand, and on the other hand its fear of high reproductivity and low assimilation of Koreans and Taiwanese, motivated the Kôseishô staff to propose some optional post-war treatments of the forced laborers, such as to dispatch them back to where they came from or relocate them elsewhere for settlement. As for the arrangement of Korean and Taiwanese soldiers in the Kôseishô's design, they were segregated from the regular Japanese army and organized into disparate Chinese and Korean military units led by Japanese officers.[37]

In the final analysis, the ambivalent interplay of practical need and identity anxiety was juggled very well, by the Japanese State and eugenicists. The interplay resulted in a pattern of concentric circles of a three-dimensional conical body in which the top center and core circle would be the emperor and his lineage, the inner circle the Yamato race as a unilinear patriarchal nation, and the outer circle pan-Asianism culturally and territorially binding together the Japanese colonies and the Greater East Asia Co-Prosperity Sphere. To this pattern of concentric circles, I would add on the eugenicists' particular design, that is, between the extensive concentric circles juxtaposing blood/culture with the parallel of quality/quantity, and with its internal/external positioning relatively shifting. The function of the eugenicists in the Kôseishô's population plan was to safeguard the boundary of self/other, interior/exterior, blood/culture, and quality/quantity of each circle line so that it would not cross over others. This was epitomized above by Furuya Yoshio's situating in the metropolis of the homeland and was represented by Nagai Hisomu's positioning in the colony in North China. In this hierarchical cone construct, the Japanese eugenicists never imagined breaking the boundary and racializing Greater East Asia into a consanguine kin entity. To put it differently, the Japanese empire of Great East Asia was "contained" virtually by the Japanese eugenicists as an ethnic body; thereby for them the concentric boundaries were clear in different levels and never contradicted one another.

In his stay throughout the wartime period as the medical school director of the Taipei Imperial University in 1937, and as the medical division director of the Peking Imperial University in 1938–45, Nagai Hisomu served as a cultural envoy to preach and convert as many as possible of the Chinese youth into pan-Asianists by appealing to them with a cultural synthesis from fragmented parts of the common past of East Asia. Nagai entitled his inaugural address, as an honorary professor at the Peking Imperial University in 1940, "Sabetsukan to byôdôkan" (The perception of discrimination and equality)[38] to justify the hierarchical cone construct of pan-Asianism and define the nature of the second Sino-Japanese War. According to Nagai, discrimination was inevitable when people were born in different races, countries and cultural institutions, while equality was an effort to erase differences. The system of feudalism in both the East and the West before the modern period had imposed suffering on the majority with its extreme development of tyranny by a minority. The rise of the French Revolution was intended to overthrow the tyranny of the upper class of aristocrats and clergy, and propagate the egalitarian idea and universal love. In a short time, this liberal new culture became decayed by capitalist logic of the survival of the fittest and laissez faire, creating a great schism between the richest and poorest, which in turn generated the reaction of communism against such cleavages. China was suffering from the dialectic oscillation of these two historical extremes, and should search for the "way of the mean" from the old wisdom of the Confucian classics. There were normally three stages of attaining wisdom: first, the confusion which mixed all things together; second, the differentiation that identified one thing from another; and third, the realization that could recognize equality from discrimination and vice versa, seeing non-wisdom as great wisdom and individual self as universal self. As such, one realized that equality and discrimination were two sides of a coin, without which one could not prevail, and that finding the way to derive a subtle mean from the two extremes would be a great life achievement.

In this speech to the Chinese students at the Peking University, Nagai said that his task in China was to form a strong middle class from the student body and thereby to bequeath the essence of Japanese Spirit and the National Polity. Figuratively, he used Mount Fuji to describe the hierarchical cone construct of the Japanese empire wherein the Japanese emperor, the Son of God, was situated at the sacred peak. Between the emperor and his retainers and imperial subjects, there existed a definite hierarchy. But the emperor treated everyone equally and benevolently as chosen people branching out from the Son of God into a Yamato clan. Love, respect, loyalty, and feeling were the natural and spontaneous expressions that bound the emperor and the subject as a unity prevailing for two thousand and six hundred years. Nagai went on elucidating the essence of the Japanese Spirit: one that emphasized self-sacrifice in the aim of protecting the composite body of the greater self including the emperor, the ancestors, the Yamato race and the nation/family (*kokka*), and was gifted with the capacity of powerful cultural assimilation.

Nagai exhorted his audience, the Chinese youth, to observe the different nature of wars in Europe and in East Asia. The invasion in Spain, Czechoslovakia and Poland by the imperialists was simply a selfish act of discrimination, greed, exploitation and inhumanity. Whites were only 20% of the world population, and yet they occupied almost 90% of the land. In contrast, before the Sino-Japanese War, Japan restrained itself on the basis of a non-expansionist principle, and tried to dissolve regional conflicts. However, Chinese warlords and financial tycoons were so deceived by their own perception of discrimination, selfish concerns and craving for power, that Japan was forced to rise with its sword to subjugate the evil warlords and *zaibatsu*. Now it was helping restore peace and prosperity to the good Chinese people and to the new order in East Asia. What Japan was doing, Nagai believed, was to provide a friend's hand and humanitarian aid out of universal love, evidenced in Nagai's own mission in China to set up a medical college and educate Chinese human resources in Peking University with Japanese funds. From the perspective of discrimination, there were things like language, culture and customs that were both similar and different between China and Japan. Nonetheless, they both had their own shortcomings and merits, and thereby had equal justification to improve each other. He encouraged his Chinese audience to discard the cultural misunderstandings grounded on discrimination, for instance sinocentrism, and meditate with a modest mind what merits they could adopt and assimilate from Japan in order to construct a new culture. Tactfully, Nagai's approach to this new culture, instead of dogmatically preaching and imposing the Japanese identity, flashed out various historical fragments evoking for the audience a common and communal past. Nagai's eloquent and enchanting speech, if not powerful enough to convert his Chinese audience to the new cultural identity and win over ready-made human resources from the side of the presumable enemy, at least contained them within a status quo.

In 1942, Nagai gave another speech, again to the Chinese students at Peking University, which addressed the problem of the quantity of the Chinese population.[39] The purport of this speech was to entrust the Chinese medical students—future physicians like Nagai himself whose ambition was to cure not merely the physical body but also the mind—with the cultural mission of pan-Asianism. Nagai's deconstructive narrative of sinocentrism and reinstatement of a new cultural synthesis of the best elements of the East and West were amazingly a simultaneous product of decolonization and recolonization, an enlightening lesson we still need to be warned against in the post-modern era. He first destroyed the glorious superficial impression of China having abundant natural resources, vast land and a huge population. China, with 25% of the world population in an area 9% of the world's geographical space, was in Nagai's eyes overpopulated. Most cultivated land was predominantly owned by landlords instead of by the vast majority of poor peasants. Wealth distribution was so lopsided that it bifurcated into two extremes of rich and poor without a middle class. Nagai went over a list of charts and statistics—along with some literary description from Pearl Buck in her Nobel Prize-winning *The Good Earth*—to demonstrate that most Chinese peasants lived under the poverty line. On top of it, there were recurring famines, wars, bandits and all kinds of disasters to plague the Chinese peasants. He asked the Chinese students to think thoroughly about the reality and significance of the misconceived impression of "China."

He earnestly urged these Chinese medical students to be superior physicians (*jôi*) to cure the problem of Chinese overpopulation. Nagai used a comparative Sino-Japanese chart of farmed area and harvest productivity, to draw a contrast between Japan's remarkable efficiency and the Chinese ill management of land and population. He also noted the nonexistence of a survey facility in China to investigate the root of the problem. As a solution, Nagai suggested institutional hardware enhancement of irrigation, fertilizer, and water dams, agricultural breeding, village credit unions and financial networks, improved transportation, local welfare and public health centers, sideline businesses, tax reduction and central control. Most importantly, though, China had to embrace the software installation of education, moral consciousness and personality enrichment, topics to which he devoted half of his speech. The specific content of ideological installation, which Nagai tried to inject into the Chinese students and thereby into the common populace, was a dialectical progression to pan-Asianism. It was meant to subjugate the vast quantity of the Chinese population without any direct confrontations and further convert them into usable human resources, that is, submissive and loyal imperial subjects.

Nagai first set up the dichotomic framework of East versus West, and subsequently those of morality/might, mind/thing, the one/the myriad, the ethical/the utilitarian, and the just rule /the military rule. He maintained that the merits of one side were the other side's demerits, and vice versa. Although Nagai admitted that the Eastern elements initially originated from China, they had nonetheless become ossified in the wake of five thousand years and remained ruins without any lively essence in the historical evidence of a dead cultural tradition. Nagai took up the Chinese internal critique of tradition to criticize China within East Asia. The long-lasting civil-service examination system turned out to be a suffocating confinement of intellectual creativity, and Confucianism became fixed in the Qing period on philological triviality. With respect to the application of knowledge, there had been no upgrade of information and revision in Chinese *materia medica* since the Tang dynasty, and gunpowder and the magnet were never developed as far as they were in the West. Accordingly, these became the demerits of the Oriental speculative idealism. After the humiliating defeats from the Opium War and the Arrow War, the Chinese self-strengthening reforms tried to imitate the Western way of might and utilitarian-oriented military rule by importing military technology and scientific knowledge. However, no sooner did the Qing court begin to deploy the reforms than the fire of the Republican revolution sparked off and shattered the already dwindling empire. The ensuing May Fourth literary revolution stamped out the remaining traditional legacy and embraced wholesale Westernization, which unfortunately was driven to the dangerous extreme of selfish individualism and liberalism that could be called a dark age of Chinese education. Before long, Sun Yat-sen's *Three Principles of People* became the educational guidelines of the Nationalist Party (KMT) and were implemented nationwide after the success of the KMT's Northern Expedition and the unification of China in 1928.

Nagai continued to attack the unbalanced and distorted development of education under Sun Yat-sen's *Three Principles of People*, which was originally intended to expel imperialism and feudalistic warlords, liberate China, treat equally every ethnicity, assert people's rights, suppress individualistic capitalism and narrow the gap between the poorest and richest. Sun Yat-sen's ethnic-nationalism in his first principle "of the people" was misused to blindly exclude foreign elements, of which the most obvious example was the anti-Japanese sentiment expressed in the textbooks and educational mobilization since the Manchuria incident. Moreover, Sun Yat-sen's two other principles, "by the people" and "for the people," ironically became the compromising foundation to reconcile Chiang Kai-shek's clique and the Communists. Ideally, Nagai respected Sun Yat-sen's pan-Asianist attempt to synthesize the merits of the East and West, as well as his stress on enlightening people. However, in reality, he saw that Sun's political ideal was betrayed by the Chiang clique and the Communists who manipulated the patriotism of Chinese youth and misled the people by the anti-Japanese campaigns. Meanwhile, the Japanese approach to the just rule and self-determination of peoples, and the search for harmony, coordination and prosperity in East Asia, were the embodiment of Sun's ideal. Using the rhetoric of *hakkô ichiu* (universal brotherhood), Nagai went on to expound the good intention and unselfishness of Japanese diplomatic treaties and each military campaign in East Asia. He encouraged his Chinese audience to take up the mission of pan-Asian co-prosperity and help direct the populace along the right path.

Nagai's narrative strategy can be best characterized by the Chinese idiom "Stealing heaven and replacing it with a rising sun (*toutian huanri*)." He first deconstructed sinocentrism and alienated his Chinese students from their Chinese past and its dialectical and reactionary present. Then he presented an alternative in Sun Yat-sen's synthesis of pan-Asianism. While retaining the form of pan-Asianism, Nagai removed the representational content presented by Chiang Kai-shek's clique and the communists, and then instilled the Japanese cone construct of pan-Asianism. The cultural tie of brotherhood and Sun's form of pan-Asianism provide a convenient vehicle to arrive at a gestalt identity, by which the Chinese still recognized their own Chineseness, and the Japanese redeployed these elements in their own imperial constellation, a concentric cone configuration extending from the Yamato race to the Japanese ethnic nation. For Nagai, this ethnic body was ready to accept the vast quantity of population within Greater East Asia as long as they had been culturally reprogrammed, or ideologically converted, through the *Kôminka* projects. However, one should not go overboard and imagine that the ethnic populations could reverse the cone hierarchy and mix up with the centered circle, since eugenicists like Furuya Yoshio stationed at the Kôseishô in the metropolis, would definitely prohibit that from happening.

CHINESE EUGENIC SOLUTIONS AND POPULATION POLICY: CONSTRUCTING AN ETHNIC NATION

If Japan did successfully control its 132 million population in the empire, as Nagai claimed above, the number would quickly add up to 346 million when combined with the occupied areas ruled by the Japanese puppet regimes in Nanking, North China and South China. These areas enclosed 17 administrative provinces and a total population of 214 million, according to the estimate of Chen Da (1892–1975, Ph.D. Columbia University, 1922), an internationally distinguished demographer and chair of the Sociology Department at the National Southwest Associated University in 1938–1946.[40] Nagai's fervor to convert the Chinese population as a strategic focus in the wartime period was precisely the focus of the Chinese eugenicist Pan Guangdan as he advocated in 1939 a campaign of spotting out the traitors or collaborators in Kunming, an area under the KMT's control, in order to secure the quantity of Chinese population. What concerned Pan the most was a fundamental sickly syndrome, whereby the vast majority of the Chinese population was susceptible to becoming Japanese collaborators. He considered this syndrome to arise from a decreasing hierarchy of four other factors: selfishness, ignorance, poverty, and sickness.[41]

He analyzed the symptom of mental and physical sickness of Chinese who were opium-addicted and thus were easily enticed by Japanese offers of heroin to either join the Japanese parades or voice support for Japanese rule. Poverty was another reason for ordinary poor people, since they became co-opted under the influence of money. Ignorance was so prevalent that people were facilely indoctrinated by Japanese propaganda and were submissively mobilized to provide military labor. Pan considered these three symptoms to give rise to numerous minor Japanese collaborators. The most serious symptom, selfishness, would cause tremendous damage, even though selfish collaborators were few in number. Among the selfish collaborators, there were some merchants and local gentry who were candid about their greedy intentions. In contrast, there were some corrupt politicians, officials, party leaders and educators who lied about their hunger for power, and who euphemized their collaboration under the noble names of "the future of national welfare" and "the everlasting peace in East Asia." Pan despised them as hypocrites who were in fact destroying China even while proclaiming themselves as China's saviors. Doubtless Pan had in mind Wang Jingwei, the well-known Chinese collaborator of the regime in Nanking. Pan admitted that to him, selfness resulted from natural selection and could not be rectified by a moral education. The other three symptoms, though, could be markedly improved through cultural environment.[42]

Pan echoed Chiang Kai-shek's campaign of spiritual mobilization in August, 1939, appealing to all gentry and educators in both the free and occupied areas to help lead local people to resist Japanese rule and manage regional autonomy and financial wellbeing. He added two basic conditions, including a minimum of material resources to meet people's basic needs, and a cleanup of corruption, without which there would remain many Japanese collaborators and submissive imperial subjects among the Chinese population, and Chiang's campaign would not achieve the goal of resistance against the Japanese.[43] Pan even encouraged those people who lived in the occupied areas to immigrate to the Great Rear areas, namely Free China in the Southwest, to rid themselves of Japanese control. Pan considered immigrants superior to local permanent residents, who were used to stability and submissively accepted what was given, that is, the Japanese puppet regime. A successful immigrant, Pan argued, possessed at least three indispensable qualities: personal drive, a persevering and risk-taking physique, and intelligent competence as he passed the test of survival. Pan enumerated various historical movements of ethnic migrations and the Han Chinese settlement from the North to the South, to support his view on the vitality of immigrants. He also cited foreign historical examples of the Anglo-Saxon immigrants' establishment of the United States, Australia, New Zealand and Canada. Most significantly, Pan commented on the Jewish immigrants expelled by the German Nazis as a great loss for Germany and a gain for the USA, because of the superior qualities of immigrants such as Albert Einstein. As Pan perceived the anti-Japanese war as a survival test, he believed that immigrants who moved to the Great Rear were born to be tough fighters for any unfavorable environment and a replenishing source for the doughty soldiers. As for those who did not migrate to the Southwest and yet joined the anti-Japanese guerrillas in the occupied areas, they also displayed the immigrants' qualities.[44]

The same bulk of Chinese population that both the Japanese and Chinese eugenicists tried to win over, played a decisive role in the Sino-Japanese war when, in the first few years, China had to fight alone the better-equipped and well-deployed Japanese army. As the distinguished Chinese general He Yingqin argued, because the firepower of one Japanese division was four times stronger than its Chinese counterpart,[45] the main resource the Chinese had to depend on was its vast population. In the initial stage of military engagement with the

Japanese army, Mao Zedong had upheld a strategy of a war of attrition. He explained that in order to confront the imperialist Japan, the semi-colonial and semi-feudal China had to overcome itself in a democratic process of reorganizing politics, the military and the economy. Most importantly, they had to achieve a far-reaching popular mobilization.[46] At the other end of the Anti-Japanese United Front, Chiang Kai-shek took a different approach to popular mobilization, until his best troops, trained by a German military advising group, were wiped out in Shanghai on August 13, 1937. The capital Nanking was subsequently lost to the Japanese army, and Chiang was forced to move his capital to Chongqing. On March 29, 1938, Chiang promulgated the Outlines for a war of resistance and national reconstruction, with which he confirmed a "mass line" approach of popular mobilization and guerrilla war tactics. This won him the first victory since the fall of Nanking in the Taierzhuang campaign on April 6, 1938. The difference between Chiang's and Mao's approaches was that whereas Mao saw the mass as the main body of his war tactics, Chiang took them as a supportive role to the main body of government and military troops.[47]

Interestingly, the Japanese collaborator Wang Jingwei, on December 18, 1939, also expressed a concern with the same bulk of the Chinese population. He pointed out that both Chiang's and Mao's guerrilla warfare tactics were a mere euphemism for restless bandits who could do nothing in a real battle but who plagued the village people. In a radio broadcast for overseas Chinese listeners, Wang stated that more than 85% of the Chinese population were peasants who barely survived from their hard work and sorrowful harvests. In addition to the regular burden of tax revenue, villagers had to put up with the perennial extortion of the guerrillas from different cliques. Wang proposed a collaborative peace movement and warned against any involution of the vicious cycle leading to a suicidal national genocide.[48]

Given the importance of the same segment of Chinese population drawn by different political forces, one might ask how large the size of Chinese population was, and why there was no initiative to devise any population control. There were no uniform computed figures of the total Chinese population even after the Nationalist Party unified China and the Republican government was reinstated in power in 1928. The highest estimate was 550 million, the lowest 300 million, and the government's own estimate was 479 million.[49] The total population had been static since then, even though there was some fluctuation depending on factors of war, famine and plague.[50]

Sun Yat-sen once complained about the problem of overpopulation, echoing the pioneering demographer Hong Liangji (1746–1809). In 1893, he submitted his patriotic complaint to Li Hongzhang, who was then in charge of modernization projects. However, Sun considered a large population to be desirable in his *Three Principles of People*. He regarded the reason China had survived the Western invasions and their colonial rule was the vast size of its population. Sheer size justified China's status as a superior nation and as a victor under the logic of the survival of the fittest. Sun opposed any policy of population control as long as Japan and the European countries were trying to increase their population. He stated that if China adopted population control, it would perish even without an imperialist invasion.[51] The Chinese high birth rate and high mortality rate canceled each other out, and led to a standstill in population growth, which worried Sun more than overpopulation. The solution Sun provided was to increase food supply from agricultural production, and stop the birth-control movement.[52] This position persisted in the Nationalist government until 1941, and their policy was to reduce the mortality rate through public-health implementation.[53]

Perhaps the only exception to the absence of population-control efforts in the Nationalist government before 1941 was Hu Dingan's. In 1934, he attempted to combine hygiene and eugenics in the national health plan when he was invited by Chen Guofu, the Jiangsu provincial governor, to take charge of the Jiangsu Provincial College of Medicine and Administration. Hu Dingan devised an educational project comprised of eight subdivisions such as racial hygiene, hygienic education, public health administration, pharmacy, regular medicine, childbirth, veterinary science, and forensic medicine. This innovative design attracted much attention from the medical commission of the Ministry of Education, and it aroused the curiosity of Japanese medical practitioners in Shanghai Medical College and in East Asia Academy in Japan.[54]

However, before the outbreak of the Sino-Japanese War, Chen's College had managed to set up only the departments of regular medicine, pharmacy, hygienic education and public-health administration. It was soon forced to halt because of war.[55] In 1938, the College was merged with the Nantong Academy into a regular national medical college named Jiangsu Medical College. When Chen's political career moved to the wartime KMT's headquarters in Chongqing, he tried to search for magic bullets from conventional Chinese herbal drug recipes. He set up a research institute of Chinese effective medicine in order to overcome the problem of wartime scarcity of medical resources. Moreover, he published and disseminated in 1942 a commonsensical pamphlet of medical knowledge, public hygiene and tips for self-treatment. This small pamphlet was an outcome of Chen's ill-fated campaign of the National Health movement, and reflected the crux of body hygiene in the ongoing New Life Movement led by Chiang Kai-shek since 1934. In the twelve outlines related to national health in the pamphlet, Chen urged that the state should intervene in individual reproductive practices, so as to prevent proliferation of genetically inferior elements and pass on the presumably good genes that were linked to the fate of the nation. Chen proposed that the State institute a central administrative organ to administer marriage affairs and promulgate eugenic regulation of premarital examinations, as well as a marriage-counseling bureau in every densely populated city to propagate eugenic ideas. The state, Chen suggested, should sponsor research projects on genetics and eugenics. Other ideas such as prenatal care, improvement of the condition of women in labor, child-care facilities, control of venereal diseases, sports, drug rehabilitation, outdoors resorts, public hygiene, mental health and nutrition were included in Chen's outlines. This pamphlet was so well circulated that the above outlines, and the eugenic message, got across in spite of the short life of Chen and Hu's campaign.[56]

We can not establish a direct causal relationship between Chen's National Health campaigns and the attempt of the Executive Ministry of the Nationalist government to organize, in March of 1941 in Kunming, a committee for population-policy research. But we can be sure that the Nationalist government was interested in instituting a sound social policy to put into practice the often-chanted wartime slogan *Kangzhan jianguo* (resist Japan and reconstruct China), which brought together people of different positions. Chen Da pointed out in his personal memoir that he was often consulted by the Nationalist government on the particular problems of population and labor.[57] Chen Da was first delegated to administer the national population-survey project in 1928, and yet the project was not thoroughly launched. In the fall of 1941, the Central Statistics Bureau of the Nationalist government invited Chen Da again to sketch a population-survey plan for Kunming and its neighboring counties in order to obtain accurate household information for conscription.

Around the same time, Chen Da was appointed as a member of the Population Policy Research Committee in the newly established Social Ministry in the Executive Yuan. The Committee included Chen's colleagues at Xinan Lianda, the eugenicist Pan Guangdan, the sociologists Li Jinghan and Wu Zelin into a Kunming team in conjunction with another team formed in Chongqing. The Committee suggested in its second meeting in March, 1942, the establishment of a marriage-counseling bureau and health-care centers in order to secure the quality of births. The Committee also recommended legislation or a decree to impose marriage restriction on mental patients and the feebleminded. At the third meeting in Kunming with the same group of specialists, the Committee decided to add several implementation plans including a population survey and statistics, and attempts to increase the frontier population, elevate living standards, prohibit abortion, and promote social life between men and women.[58] In the meantime, the Kunming team convened twice with the one in Chongqing in order to reach a consensus for the guidelines on population policy drafted mainly by Chen Da and Pan Guangdan. In the two joint meetings with the Chongqing team, there were some disputes between the Committee and the Nationalist party members over the guidelines. These disputes included: 1. Some party members insisted dogmatically on their understanding of Sun Yat-sen's strategic promotion of population growth as a permanent unlimited expansion. Chen Da suggested the compromising term of "rational increase" to settle this dispute. 2. Based on the above misunderstanding, some party members disagreed with the birth-control plan; therefore, the term "birth control" would not appear in the guideline, but it was incorporated in the implementation blueprint. 3. In the Fifth National Assembly of the Nationalist Party, one senior party member objected to the stipulation of sexual education, but the Committee insisted on its fundamental necessity and importance in the plan. As a result, the stipulation was merely modified by different wording. On May 5, 1945, the Guideline of National Population Policy (*minzu baoyu zhengce gangling*) was passed in the Sixth National Assembly of the Nationalist Party.[59]

In this guideline, the eugenics principle of preventing the proliferation of defective elements was eventually incorporated. In order to safeguard eugenic quality, the methods of premarital examination, venereal-disease prevention, and the separation or sterilization of hereditarily defective people were adopted. Comparing this guideline with Chen Guofu's twelve outlines, there was not much difference, except that the former increased several articles such as the prohibition of abortion, infanticide, concubinage, abduction and human trafficking; helping increase the frontier population by implementing public health, educational facilities, industrialization and environmental improvement, and most intriguingly by encouraging interracial and interethnic marriage in order to strengthen national unity and provide human resources for military conscription.[60] This particular feature of interethnic and interracial marriage encouragement formed a sharp contrast with the Japanese population policy. For the eugenicist Pan Guangdan and his sociologist colleagues, the anti-Japanese war, like its numerous historical precedents of war, drove a large-scale migration of the Han population to the West and Southwest that would rejuvenate racial vitality or, in Pan's own terms, "racial stamina" of the Han Chinese.

In spite of his practical concern with the ethnic minorities as a source for military conscription though the cultural device of sinification and the biological encouragement of intermarriage, theoretically Pan did not take the Hans and minorities as a single whole for the judgment of racial quality. He pinpointed in 1939 and later in 1943 in his lectures to the Eleventh Military Division stationed in Dali, Yunnan, that the concept of "race (*zhongzu*)" was often mistaken and intermingled with two disparate levels of significance of species (*zhong*). One was the taxonomical distinction among species and their different characters in terms of kinship, and the other was the eugenic distinction of lineage and its genetic transmission, on which one had to impose a value judgment of genetic quality in order to avoid degeneration. Superior and inferior lineages, according to Pan, could both be located in the same racial group, or in every racial group of an ethnic nation. Pan critiqued Hitler's anti-Jewish racial discrimination as an arbitrary outcome of such a mistaken confusion. Nordics and Jews could not count as different species because they belonged to humankind in the same species, and each of them was a combination of different racial groups that reproduced both superior and inferior lineages. The Nazis arbitrarily assigned to Jews the inferior position, and conflated a double mistake into a misguided population policy. Pan contested that if the Nordic race was a superior one, as it claimed, why bother to impose a sterilization policy in their eugenics movement? The Nazi method of trying to push a refined speciation of the Nordic race would not do itself any good, because diversity was also indispensable in order to retain the adaptive ability in case of drastic environmental change. From the viewpoint of the evolutionary process, Pan said, speciation and diversification were the dialectical ways in which species and individuals came into being, and precisely because of this, ignoring one way or other would make the species extinct.[61]

Pan once alluded that China was young as an ethnic-nation, based on the study of Griffith Taylor's *Environment and Race*, because it kept intermingling with other young races through different historical periods. Likewise, the adaptive strategy for an inviable and new environment would be to rebuild the genetic makeup through racial intermarriage, as when the Hans migrated to the West and Southwest along with the Nationalist government retreating into the Greater Rear, and the hinterland inhabited by Mongols, Muslim Turks, Tibetans, and Uygur in Xinjiang. According to Pan, this wartime migration would bring the racial sensibility that the frontier's ethnic minorities were genetically superior to their own culture in this environment. In his lecture to the eleventh military division stationed in Yunnan in 1943, Pan suggested that the State carefully "elevate the cultural standard of the minorities" without downgrading their vigorous genetic quality, which the overly civilized and thereby degenerated Hans were lacking and seeking to supplement. Using Confucius' concepts of *wen* (civilized) and *zhi* (unadorned) as the two mutually supplementary aspects of a gentleman, Pan wished that the Hans and the frontier minorities, like *wen* and *zhi*, could supplement each other in order to reproduce better Chinese offspring.[62]

CONCLUSION: POSTWAR REFLECTIONS OF TWO EUGENICISTS

After the eight-year Sino-Japanese War, the Japanese eugenicist Nagai Hisomu was expatriated back to Tokyo. Meanwhile, Pan Guangdan returned to Tsinghua University in Beijing when the Xinan Lianda, the wartime associated Southwest University, was disbanded and its faculty returned to their prewar institutions. Whereas Japan was occupied temporarily by the United States in 1945–52, China was plagued again by the civil war between the Nationalist party (KMT) and the Chinese Communist party (CCP) from 1945 to 1949, when Chiang Kai-shek's troops were defeated and the Nationalist regime was forced to retreat to Taiwan. In Japan, the title of the National Eugenics Bill was changed into the Eugenics Protective Bill in 1948. Between 1941 and 1945, 192 men and 243 women received sterilization surgery based on the Kôseishô's Eugenics Bill, while it was estimated that 8,820 men and 6,399 women were yet to be sterilized on the waiting list. In China, since the domestic turmoil of the civil war and the Nationalist regime's relocation to Taiwan in 1949, the implementation and legal enforcement of the Guideline of National Population Policy was disrupted.

It seems ironic that the eugenicists returned to their original and more autonomous position prior to the War, and condemned war as a dysgenic mechanism. After returning to Japan in 1945, Nagai reflected in 1946 on the failure of the War, which he regarded as a divine punishment caused by several ambitious warlords who ignored the scientific spirit and depended on the suicidal Kamikaze spirit. In the long-enduring war of attrition, the Japanese warlords suicidally dispatched all the best stock of Japanese soldiers to the battlefront, leaving the weak and inferior ones at home, where they married and unexpectedly procreated problematical offspring. Consequently, the War, in Nagai's opinion, had promoted socially deleterious selection in the Japanese population. It would take a long time and much striving to reverse this undesirable tendency, which was once again in the line of duty of the Racial Hygiene Association and the self-delegated mission of the eugenicists.[63] The scientific spirit Nagai reflected on, as analyzed in the above section, was a cultural synthesis of the Japanese Spirit and Japanese-characterized science. He ascribed the war failure to a selfishness of ambitious warlords without conforming to the self-effacement of the synthesized "scientific spirit." In other words, what Nagai lamented was that the "scientific spirit" was not pushed and extended far enough to bind everyone in the nation.

As for Pan Guangdan, he transformed his wartime sympathetic attitude toward ethnic minorities, and eugenic interest in understanding the genetic makeup of the Han Chinese as a grand racial and ethnic fusion, into a cultural critique of Sinocentric nationalism or Han chauvinism. He saw nationalism and imperialism as two sides of one coin which had been epitomized in racial supremacy by the Japanese empire and Hitler's fascist expansion. He reflected that although China was not able to become an imperialist power, it was nonetheless ingrained with Han chauvinism expressed in every historical treatment of ethnic minorities. Traditionally, expressions of Han chauvinism included military conquest and suppression, and cultural sinification that bound ethnic tribal kingdoms with feudal kinship, the Confucian five ethics, self-delegated moral superiority and a tributary system. After the Republican Revolution, Sun Yat-sen's ideal of ethnic autonomy and peaceful coexistence was never materialized by the Republican government. Pan critiqued that the Republican government, even though changing to a different name for administration, followed de facto their historical precedent of politically subjugating and culturally dominating ethnic minorities. This unfulfilled ideal of ethnic equality and autonomy, Pan believed, awaited the liberating force of the People's Revolution postulated in Maoism and Lenin-Marxism. In 1952, Pan switched from the discipline of sociology at Tsinghua University to the Central College of Ethnology, where he devoted his career to the study of minority cultures. He was politically accused as a rightist reactionary by Meng Ming and Huang Dongsen in 1957 for his earlier promotion of eugenics via affirmative recognition of the traditional selection mechanisms of filial piety, familism, matchmaking of equal socio-economic standing, and an examination system. The chief reason for the attack on Pan was his opposition to Lysenkoism. In 1939 Pan had rejected the theories of genetic inheritance of acquired characters promoted by Paul Kammerer (a German geneticist invited to Russia for research) and Ivan Pavlov under the dogmatic influence of

Stalinism.[64] In 1967, at the height of the Cultural Revolution, Pan was tortured to death because of his politically incorrect theories.

NOTES

[1] Herbert Spencer, *On Social Evolution—Selected Writings*, ed. by J. D. Peel, (Chicago: University of Chicago Press, 1972), pp. 168–169. This selection was excerpted partially from Spencer's various works, *Social Statics, The Study of Sociology* and *Principles of Sociology*.

[2] Ploetz was taken seriously enough to be once nominated for a Nobel Prize. His eugenics ideas were often appropriated by Hitler and the Japanese eugenicists as well. See Robert N. Proctor, *Racial Hygiene—Medicine under Nazi*, (Cambridge: Harvard University Press, 1988), pp. 28–29.

[3] This evolutionary idea of the war of total mobilization was especially pronounced by the publication endorsed by the Japanese Konoe Cabinet, *Kokka sodoin* (National mobilization), (Tokyo: Kokufukai, 1937), pp. 86–92.

[4] See Matsumura's article, "Minzoku no kakusei" originally published in *Bunge shunjû*, (1933), no. 11 and the critique of Matusmura's article by an anonymous Marxist biologist has been included in the collection *Nippon kagaku gizyutusi taike*, (Tokyo: Daiichi hoki shupan kabushiki kaisha, 1964–72), vol. 15, pp. 413–415. Komai's article, "Dai-Tôa sensô to idengaku (The Greater East Asia War and Genetics)" and Tanaka's "Sensô to idengaku (War and Genetics)" also included in the same volume, pp. 409–411.

[5] Pan Guangdan, "Kangzhan de minzu yiyi" in his *Yousheng yu kangzhan* (Eugenics and the anti-Japanese war), (Chongqing: Shangwu shuju, 1944), pp. 67–72.

[6] See *Kôseishô nijûnen shi*, edited and published by the Kôseishô in 1960, pp. 94–95.

[7] Ibid., pp. 96–109.

[8] John W. Dower, *War Without Mercy*, (New York: Pantheon Books, 1986), pp. 270–1.

[9] Nagai Hisomu, "Ketsu to bunka to nihon minzoku" in *Inochi*, (October, 1937), pp. 202–221.

[10] Ibid., p. 220. Jon Davidann's article "Citadels Of Civilization: US-Japanese Relations in the Interwar Period" in American Influence in Asia: Transpacific Relations in the Twentieth Century (forthcoming April, 2002) also deals with the same argument the Japanese made about their civilization.

[11] Despite a series of debates among historians and ethnologists on the origin of the Yamato race and chronology, the unfolding of Japan basically started from the imperial family (tribe) from the North who unified the tribes in the South and established the Yamato supremacy. While winning victories over others, the Yamato uniquely combined the best characteristics of militaristic spirit (*bu*) from the North/Mongol and civilization (*bun*) from the South/China, and assimilated all communities of the archipelago into the same race. For details, see Stefan Tanaka, *Japan's Orient: Rendering Pasts into History* (Berkeley: University of California Press, 1993), chapter IV, pp. 153–187.

[12] Chou Wan-yao, "The Kôminka Movement in Taiwan and Korea: Comparisons and Interpretations" in *The Japanese Wartime Empire, 1931–1945*, ed. Peter Duus et al. (Princeton: Princeton University Press, 1996), pp. 40–68.
[13] Oguma Eiji, *Tan'itsu minzoku shinwa no kigen*, pp. 235–270. From Oguma's study of the entangled historical events of the *Kôminka* movement and eugenics movement, I hope the racism teased out from the colonial science and policy would demythologize the nostalgic romanticization of Japanese colonial rule in post-colonial Taiwan.
[14] Ibid., pp. 242–246.
[15] See Kirihara Shin'ichi, "Kesseigaku kara mita waga kuni to kono shûi minzoku (The Japanese and the surrounding races seen from serology)" in *Yûseigaku*, vol. 2 (1925), no. 11, pp. 13–14. Kirihara was the medical officer working for the Government-General of Korea hospital.
[16] See Shirayanagi Shûko, *Nihon minzoku ron*, (Tokyo: Chikura shobô, 1942), pp. 6–7. Shirayanagi was a historian and early socialist; he was tried as a war criminal in the postwar period.
[17] These principles are: 1) blood type is inherited; 2) blood type would never change in one's lifetime even under the influence of food, endemic factors and diseases; 3) the distribution of blood type, even after several generations' interbreeding, remains constant. See Furuhata Tanemoto, "Ketsuekigata yori mita nihonjin (The Japanese seen from the blood types)" in *Nihon minzoku*, ed. by the Tokyo Anthropological Association, (Tokyo: Iwanami shoten, 1935), pp. 85–101, especially p. 95.
[18] Ibid. This specific Japanese pattern that Furuhata derived from 301,959 samples was Blood Type O 30.5%, Type A 38.2%, Type B 21.9% and Type AB 9.4%.
[19] Ibid.
[20] Cit. note 13, p. 250.
[21] Ikeda Shigenori, *Shidôsha minzoku no yuseigaku teki ishin*, (Tokyo: Nihon shuppansha, 1942), pp. 3–5.
[22] See Bunzo Hashikawa, "Japanese Perspectives on Asia: From Dissociation to Coprosperity" in *The Chinese and Japanese: Essays in Political and Cultural Interactions*, ed. by Akira Irie (Princeton: Princeton University, 1980), pp. 328–355. Also Yamada Shôji, "Jiyû minkenki ni okeru kô-a-ron to datsu-a-ron: Ajia shugi no keisei o megutte" in *Chôsenshi kenkyû kenkyûkai ronbunshû*, vol. 6, 1969, pp. 40–63.
[23] This concern was expressed in an editorial pledge in the *Yomiuri shinbun* (Yomiuri Newspaper) on July 4, 1939, and this editorial article was included in the *Kôseishô nijunen shi*, p. 222.
[24] Ibid.; the Outlines are listed in pp. 216–218.
[25] See the *Shokuin roku* 1940 (Staff directory), published by the Cabinet, pp.182–191.
[26] Furuya Yoshio, "Minzoku kokusaku no shomondai (The problems of national policy of race/ethnicity)," in *Yûseigaka*, vol. 16 (1939), no. 189, pp. 3–13, and no. 190, pp. 1–10.
[27] Ibid.
[28] Ibid.

[29] Furuya Yoshio, *Kokudo, jinkô, ketsueki*, (Tokyo: Asahi shimbunsha, 1941), pp. 176–177.
[30] Ibid., p. 178.
[31] Ibid., pp. 179–181.
[32] Ibid., pp. 182–183.
[33] Even though the drafter of this secret document was not identified, it can be seen as a team production endorsed by the eugenicists staffed in the Kôseishô.
[34] Cit. note 35, pp. 253–254.
[35] Michael Weiner, *Race and Migration in Imperial Japan: the Limits of Assimilation* (London: Routledge, 1994), p. 193.
[36] Chou Wan-yao, "The *Kôminka* Movement," p. 65.
[37] Cit. note 35, pp. 255–256.
[38] Nagai's lecture was published in *Tôa renmei* (East Asian League), vol. 2, no. 2 (Feb., 1940), pp. 90–98.
[39] This second lecture "Pekin daigakusei ni tsugu" was published in *Chûô kôron*, no. 654 (Feb., 1942), pp. 54–66.
[40] According to Chen Da, it only 5% of the total population who migrated to the Southwest; see Long Guanghai, *Zhongguo renkou* (Taipei: Zhonghua wenhua, 1955), p. 185.
[41] Pan, *Yousheng yu kangzhan*, pp. 110–113. Pan had discussed in 1934 these four common symptoms of the Chinese populace in his article "Minzu bingxiang (National sickness)."
[42] Ibid.
[43] Ibid, pp. 114–119.
[44] Ibid, pp. 79–86.
[45] Ishijima Noriyuki, "Kô-Nichi sensô" in *Taikei: Nihon gendaishi* (Tokyo: Nihon hyôronsha, 1979), vol. 2, pp. 186–231, especially, p. 187.
[46] Ibid., p. 188.
[47] Ibid., pp. 198–200 and pp. 208–209.
[48] Wang Jingwei, "Jinggao haiwai qiaobao" in *Wang Jingwei xiansheng heping yundong yanlunji* (Guangzhou: Zhongshan ribao, 1939), pp. 15–17.
[49] Zhang Jingyuan, *Zhongguo renkou wenti* (Taipei: Zhongguo renkou xuehui, 1959), p. 113.
[50] See the charts of various estimates from 1911 to 1948, and the rough totals in every dynasty and their subsequent sources prepared by Long Guanhai in *Zhongguo renkou* (Taipei: Zhonghua wenhua, 1955), pp. 166–168.
[51] Sun Yat-sen, *Sanming zhuyi*, the second lecture of "minzu zhuyi," (Taipei: Zhengzhong shuju, 1978), pp. 15–16.
[52] Zhang Jingyuan, *Zhongguo renkou wenti*, p. 303.
[53] For a study of the Nationalist implementation of public health, see Ka-che Yip's *Health and National Reconstruction in Nationalist China: The Development of Modern Health Services, 1928–1937* (Ann Arbor: AAS monograph and occasional paper series, no. 50).
[54] Ibid., pp. 510–519.
[55] Another reason was that because of its amalgamation of traditional Chinese medicine and modern Western medicine, the Ministry of Education was not able

to assign it to a single predesigned pigeonhole for registration. Ibid., pp. 501–504.

[56] Chen emphasized that these ideas in the outlines were well known by many people by the time of publication of this pamphlet; see his *Weisheng zhidao,* in the collected work *Chen Guofu xiansheng qianji* (Taipei: Zhengzhong shuju, 1952), vol. 6, pp. 157–248, especially pp. 233–244.

[57] See Chen Da's personal memoir, *Langji shinian* (Chongqing: Shangwu, 1946), p. 379.

[58] Ibid., pp. 443–444.

[59] Ibid., p. 445.

[60] Long Guanghai, *Zhongguo renkou,* appendix, pp. 1–7.

[61] Pan, *Yousheng yu kangzhan,* pp. 35–46.

[62] Pan, "Kangzhan yu minzu" in *Pan Guangdan minzu yanjiu wenji* (Beijing: Minzu chubanshe, 1995), pp. 106–110.

[63] Suzuki Zenji, *Nihon no yûseigaku,* pp. 167–168.

[64] Pan, *Yousheng yu kangzhan,* pp, 41–42.

6

CONCLUSION: WHAT CAN WE LEARN FROM EUGENICS?

> The ruler of the southern sea was called Shu. The ruler of the northern sea was called Hu. The ruler of the central zone was called Hundun. Shu and Hu often met on Hundun's territory and being always well treated by him, determined to repay his kindness. They said, "All men have seven holes, for seeing, hearing, eating and breathing. Hundun alone has none. We will bore some for him." So every day they bored one hole; but on the seventh day Hundun died. (*Zhuangzi*).[1]

> Just as it would be perverse to forbid all food because someone died from rancid food, or forbid all boats because someone died from riding in a boat, so too would it be perverse to abolish all weapons simply because someone lost his state using weapons. The reason weapons cannot be abolished is analogous to the reason why water and fire cannot be abolished. If a person uses them skillfully, he enjoys good fortune; if he uses them incapably, he suffers calamity. It is just the same as using herbal medicines: if you procure a good herb, you can help another to survive; but if you procure a poisonous herb, you may kill someone. Righteous weapons are the world's good herbs, but they are even more important. (*The Annals of Lü Buwei*)[2]

EUGENICS IS A TERM LOADED WITH CONTROVERSIES AND EMOTIONS. IT HAS been a very popular topic between 1880 and 1945, similar to the current heated discussion of heredity manipulation on designer babies, gene therapy, stem-cell research and human cloning. Given the multifold signification of eugenics in the history as I have shown so far, we learn that the reason it never disappears is because it is a well-intended subject speaking to a human desire for disease free and healthy life. No eugenics advocates propagated the ideas thinking they were evil and pseudo-scientific and would do harm. Just like the Hundun episode in the first epigraph, Shu and Hu meant well to develop Hundun's seven senses so as to enjoy life better. Unfortunately, Shu and Hu's good intention resulted in Hundun's death. It is clear that Shu and Hu have a specific definition of a better way of being, and Hundun's is inadequate by their standard. Even in the most negative version of eugenics in Nazi medicine, as Robert N. Proctor recently shows us, their concerns and programs for better health by no means differed from ours nowadays.[3] Proctor's scholarly finding is not to exonerate Nazi atrocities of euthanasia, human experimentation and mass murder in the concentration camps, and hence beatify eugenics. Instead, as he wrote first in 1988 of Nazi crimes and later of Nazi cancer studies, Proctor is directing us to its very complexity and warning us that if we do not study carefully the process toward unintended consequences, we might as well repeat the same mistakes. Moreover, an exclusively negative presentation and fixed definition of eugenics would only reduce our sense of crisisand mislead us to either drop the stigmatized subject or overlook the complicity of science resurfacing and refashioning in different contexts.

As the second epigraph argues, we should not curb our pursuit of a better life and good medicine simply because it was abused before. I agree with Daniel Wikler's summary of eugenics lessons: 1) Eugenics is a valuable case study which demonstrates how the prestige of science can be used to disguise the moral premises and motives for a social movement, and how class, racial, and other biases can exert powerful and damaging influence over such a movement while remaining virtually invisible to its advocates. And it is another illustration of the sad thesis that good intentions can lead to evil consequences. 2) Eugenics offers a perspective on the practices of our own era. Critics of certain practices in clinical genetics, and of some contemplated uses for the genetic technology of the future, maintain that these are eugenics in disguise. We must be able to evaluate this claim so that we can avoid the errors and wrongs of the past as we frame public policies for genetics in the future. 3) We should avoid an unthinking rejection of every eugenic thought or value. The fact that eugenicists were in favor of a particular measure or goal is not in itself sufficient reason to oppose it (because we all live within certain categories of inclusion and exclusion). We need a good analysis of which eugenic aims were wrong-headed and why. We might judge that some of the questions to which eugenicists proposed answers ought not be ignored.[4]

Specifically, my comparative study of Japanese and Chinese eugenics as two possible outcomes shows the complexity of the issues and escapes the indictment of either vindicating the bad name of eugenics in East Asia before 1945 and defending the current implementation of a eugenics bill in China, or ignoring the local desire for better health and better life. Pan Guangdan's 1930s eugenics proposition of non-Sinocentric miscegenation, famine relief, elevation of economic productivity, impartial distribution of wealth, control of urbanization, family planning, and freedom of speech, press and political organizations is still a durable answer to many current problems of Chinese wellbeing.

The different eugenic approaches of Marion Yang's birth control and Yoshioka Yayoi's anti-abortion and birth control speak to the core of women's perennial suffering within the regime of patriarchy. I privilege their voices by allowing them space for biographical details, because they are understudied and underrepresented in relation to their male counterparts. The same privileging position would apply to subaltern women should details about them be revealed by a future ethnographical study. From a personal and gendered standing, the eugenics bills, Japanese and Chinese, provide women a long-deserved package deal of motherhood protection and preliminary prevention of sexual transmitted diseases.

The Japanese silence about fascist atrocities and the eugenic past of racial exclusion aggregates messier the entanglement of complicity of science and subjectivity in general. Perhaps, the Japanese state cannot handle the political consequences and possible financial liabilities should they open up this traumatized history. Perhaps, it is still unable to come to terms with the paradox of modernity and the disparity between means and ends, such that good intentions could lead to disasters, and high-minded intellectuals and nonconformist cultural critics could become the accomplices of the regime. But, then, are we able to?

NOTES

[1] This paragraph is excerpted from the chapter Yingdiwang from Zhuangzi, trans. by Herbert A. Giles, (London: George Alien, 1923), pp. 90–91.
[2] Excerpts rom the chapter of Dangbing (Rattling Weapons), trans. by John Knoblock and Jeffrey Riegel, (Stanford: Stanford University Press, 2000), p. 177.
[3] Robert N. Proctor, *The Nazi War on Cancer*, (Princeton: Princeton University, 1999).
[4] Daniel Wikler, "Can we learn from eugenics?" in *Journal of Medical Ethics*, (1999), no. 25, pp. 183–194.

7

EPILOGUE: THE COMEBACK OF EUGENICS DISCOURSE

AFTER THE JAPANESE DEFEAT IN 1945, THE JAPANESE NATIONAL EUGENICS BILL (*Kokumin yûseihô*) of 1940 was changed into the Eugenic Protection Law (*Yûsei hogohô*) of 1948. The Law had attenuated emphasis on motherhood protection (*bosei hogo*), in order to tackle the chaotic postwar problem of unwanted pregnancies, due to prevalent adultery and interracial sexual affairs between American soldiers and Japanese women, and to the illicit resort to abortion.[1] This law has stood without further amendment since 1952.[2] According to the law, there are five justifiable grounds for voluntary sterilization: 1) when the person or the spouse are suffering from hereditary diseases, mental illness, deformity or feeblemindedness; 2) when any blood relative (within the fourth degree) of the person or the spouse are suffering from hereditary diseases, mental illness, deformity or feeblemindedness; 3) when the person or the spouse has contracted leprosy; 4) when the continuation of the pregnancy or childbirth would be detrimental to the health of the mother; and 5) when childbirth would lower significantly the health of a mother who already has several children.[3]

On the basis of this law, in the decade 1950–1959, there were 321,342 cases of sterilization for the reason of maternal health protection, as opposed to 3,353 for hereditary diseases, 1,135 for leprosy, 648 for nonhereditary mental illness and 8,408 cases of compulsory sterilization. For pregnancy termination, there were 9,762,093 cases for maternal health protection, 30,318 for hereditary diseases, 5,112 for leprosy and 8,324 for rape and forced adultery.[4] Evidently, sterilization and pregnancy termination were performed predominately on the grounds of maternal health protection. As a result, we can ascertain that the nature of Japanese postwar eugenics has been determined by the practice of maternal health protection.

In comparison, the Chinese eugenics bill within the Guideline of National Population Policy (*minzu baoyu zhengce gangling*) was passed on May 5, 1945, but was left aside when the Nationalist government fled to Taiwan and the communist regime took over mainland China. However, the eugenic concern did not disappear and was actually pre-existing in the marriage regulations in the Border Area Governments under Communist control during the war. For instance, 1931 marriage regulations in the Chinese Soviet Republic stated that persons suffering venereal disease, leprosy, tuberculosis, mental disease or paralysis were forbidden to marry, unless it was considered permissible after medical examination.[5] The same regulations were carried on through the 1934 marriage law, the 1939 marriage regulations of the Shensi, Kangsu, Minghsia Border Area, and other revised provisional marriage regulations.[6] In 1950, the National Marriage Law stipulated marriage prohibition if one party suffered from venereal disease, mental disorder, leprosy or any other disease which was regarded by medical science as rendering a person unfit for marriage.[7] The article has since left a lot of space for individual interpretation by different provinces and municipalities, and the relevant judicial and legislative organs have not issued any appropriate interpretation.

From the late 1970s, China has had to confront the pressing problem of population growth and the limits of its resources. In 1979, the implementation of the "one-child" policy mandated that couples have no more than one child in urban areas, while families in rural areas may have up to two children in very limited circumstances. Since then, eugenics has been tied closely with population control in order to achieve the goal of fewer but healthier babies with the prevention of birth and genetic defects. In 1980, the People's Daily published an article that gave a list of people who were considered unproductive and a danger to society, and who subsequently should not be allowed to breed, including imbeciles, hemophiliacs and the color blind.[8] Thanks to political maneuvering on the part of the population-control program, the "names" of eugenics veterans such as Pan Guangdan and Marion Yang (Yang Chongrui) were able to be cleansed from the anti-Rightist Movement purge, and re-surfaced in the eugenics discourse in the 1980s.

However, in the comeback of eugenics discourse, there was no history of Pan's and Yang's activities, or related discussions of their historical significance, since the Republican era. Examining the eugenics literature of the 80s, the course of eugenics development seemed to start, as always, with Francis Galton's coinage and heredity studies, and then it was as if nothing happened for a century, except Hitler's brutality over the Jews, and a sudden emergence of the 80s campaigns to enlist national support for improvement of population quality and the national/racial prosperity.

One of the articles collected in the title *Xingjiaoyu yu yousheng* (Sexual education and eugenics), published in 1985, pled for national legislation of the eugenics bill, since some of the measures were socially compulsory and would not be enforced without laws. This article stated that the first eugenics bill was enacted in 1909 in Indiana, the United States, and that countries like the United States, Denmark, Sweden, Norway, Finland, Iceland, Switzerland, Canada, Mexico and Japan had enacted eugenics bills. China had not, except for the marriage regulations to restrain carriers of hereditary diseases and leprosy from marriage. In addition to marriage prohibition, the author pled for a eugenics-bill enactment to entitle the State to mandate compulsory sterilization on individuals classified as medically unfit. Marriage prohibition worked only on those who would directly endanger their spouses through the sexual relationship, whereas sterilization sought to prevent jeopardy in the next generation.

The author encouraged emulation of the Japanese Eugenic Protection Law for the sake of the public interest and the prevention of transmission of hereditary diseases. According to the Japanese Eugenic Protection Law, these hereditary diseases included schizophrenia, manic depression, psychosis, epilepsy, mental retardation, myotonic dystrophy, amyotrophic lateral sclerosis, Huntington's chorea, hemophilia, color-blindness and the like. The author suggested using the above category of congenital and hereditary diseases as the basis for further research, in order to determine what diseases should be considered for compulsory sterilization in China. The author also suggested that the eugenics bill include other regulations such as the performance of euthanasia, with the consent of parents or at the suggestion of medical professionals, on newborns in the case of congenital diseases or other serious deformity not detected in prenatal tests. The author considered this measure as humanitarian, and argued that it would spare unnecessary burden for the family, society and the nation, and guarantee every couple a smart and healthy child.[9]

In 1986, the Ministry of Public Health and the Ministry of Civil Affairs put out a Circular Concerning Premarital Medical Check-ups, which included three categories affecting marriage. "Marriage is prohibited between close relatives and between mentally retarded people. Marriage is to be postponed when one or both parties are suffering from schizophrenia, manic-depression or other psychoses. Marriage is permitted but childbirth forbidden [when] either party whose inherited disease, such as schizophrenia, manic-depressive psychosis, or other types of psychosis as well as congenital heart disease, is in a stable condition."[10] This Circular served as a preliminary for the later drafted Eugenics Law, and an experimental blueprint. Gansu province was chosen as a testing ground because it is one of the poorest provinces and is well known for its large population of people with learning disabilities. The Circular in Gansu ruled that intellectually impaired persons considering marriage must be sterilized, and if already married but pregnant must have an abortion. It has been reported that 1000 women were sterilized in the first year of the implementation.[11]

Facing the financial burden of 10 million people with learning disabilities and another 10 million people disabled from birth, the Ministry of Public Health issued a draft Eugenics Law in December, 1993, which was aimed at "preventing new births of inferior quality," particularly in underdeveloped and economically poor areas. Restrictions on marriage and childbirth are to apply to those with hereditary, venereal or reproductive ailments, severe psychoses or contagious diseases. Despite criticism and protests in the mass media in the West, the law was promulgated on October 27, 1994, and took effect in June of 1995. It is important to note that while the English name of the draft was changed to the Maternal and Infant Health Care Law, the Chinese name of Eugenics Law remains unchanged.[12]

The Maternal and Infant Health Care Law requires physicians to recommend that couples should postpone their marriage if either one is found to have an infectious, contagious disease or an active mental disorder. If one partner is diagnosed with a serious hereditary disease, the couple may marry only if they agree to use long-term contraception or to give up childbearing by undergoing ligation. China received trenchant criticism and attacks from Human Rights activists, and threats to boycott the 1998 International Congress of Genetics in Beijing from the U.K.'s Genetical Society and some genetics societies in other countries.[13] Under such media uproar and international pressure, the Ministry of Public Health and the Legislative Affairs Commission of the National People's Congress issued an explanation stating that sterilization should be allowed only with the subject's consent. It made this announcement to scientists attending the International Congress of Genetics in Beijing on August 3, 1998. This explanation has the force of law until the National People's Congress formally revises the original statute. The explanation also makes clear that people who have tested positive for the human immunodeficiency virus (HIV), believed to cause AIDS, do not require permission to marry; however, patients diagnosed as suffering from full-blown AIDS in China are still banned from marriage.[14] From the Chinese news media, it is interesting to point out that in the process of promulgating and revising the Law, there are arising in China new scientific professionals, called bioethicists, engaged in the revision of legislation. Their most vocal representative is Qiu Renzong, the director of bioethics at the Chinese Institute of Philosophy in the Chinese Academy of Social Sciences.

Reviewing the above postwar development of Chinese eugenics, and comparing it with its historical progression and fate in the Republican era, China seems to have achieved now what it was not able to do in the prewar period, with its successful emulation of the Japanese model, that is, the State as a strong oversized head ruling the national body. Only now, China is still facing the same weakening body. Why was history not learned? Why was there no historical account of Chinese eugenics development when Chinese eugenicists pushed for reinstatement of a eugenics bill in the 80's, and why did the Chinese eugenicists encourage emulation of the Japanese revised eugenics bill within the eugenics discourse? Comparing the nonsynchronic globalization of biomedical technology administered by the Rockefeller Foundation almost a century ago, what is the significance of today's synchronic development of globalization exemplified by the rapidly growing field of bioethics as a new profession, and the recurrent eugenics rhetoric in the discussion of the human-genome project and with respect to reproductive rights? Likewise, what is the meaning of today's globalization when the annual National Eugenic Science Meeting in China has become a network for the multinational enterprise of circulating advanced genetics technology and equipment?

These questions provide me leverage to move beyond the scope of this book, and an opportunity to review the subject from a different perspective for comparison. I do not have easy and immediate answers, merely a few proposed topics for further exploration. First, in order to locate specifically the nonsynchronic nature of globalization decades ago, I would like to conduct a comparative study of genetics and eugenics research done under the patronage of the Rockefeller Foundation in a global context from the turn of the century to the end of WWII. This is also to further answer the intriguing question I posed in the third chapter—if the Chinese genetic establishment was borrowed from the American scientific model, which was shaped especially under the auspices of the Rockefeller Foundation, why was it that Chinese geneticists, as recipients of RF grants, did not follow their American geneticist counterparts, as many as half of whom joined the eugenics campaign? A global measurement of the RF sponsorship and financial distribution of biomedical research at home and abroad should detect a hierarchy of research priorities in terms of geographical distance and ethnic affinity, and will illuminate the nonsynchronic mechanism in the globalization of biomedical technology.

Furthermore, in view of recent startling developments in genetics technology and the contemporary meaning of eugenics being intriguingly mixed and elusive as we observe its postwar development, I hope an ethnographic study of the intricacy of social relations of the members of this developing eugenics network will help tease out its global significance, and give a balanced account of its historical sensibility and future plausibility. The displacing confusion of eugenics meanings might be due to the fact that we are actually faced by two kinds of temporalities of globalization, one being historically deposited and another horizontally disseminating.

The next proposed topic would be the postwar debates on Lysenkoism in China and Japan, and the scientists' positioning in this controversy in the backdrop of shaping the dichotomous confrontation and politics in the Cold War era. This study proposes in part to fill the historical gap that, when the Chinese eugenicists pushed for reinstatement of a eugenics bill in the 80's, there was no historical account of Chinese eugenics development. It also proposes to resolve the puzzle of why the Japanese revised eugenics bill became a model for emulation within the Chinese eugenics discourse.

NOTES

[1] See Yoshimasu Shufû et al., *Yûseigaku* (Eugenics) (Tokyo: Nankôdô, 1961), pp. 188–196.

[2] See Sandra Buckley, "Body Politics: Abortion Law Reform" in *The Japanese Trajectory: Modernization and Beyond,* Gavan McCormack and Yoshio Sugimoto, ed. (Cambridge: Cambridge University Press, 1988), pp. 205–217.

[3] Yoshimasu, *Yûseigaku,* p. 189 and p. 191. The grounds for pregnancy termination differed from those for sterilization in the fourth point by adding the economic reason, and in the fifth point that when the pregnancy is induced by menace or violence or when it occurs in a woman forced into adultery without opportunity to resist or reject. Also, see Buckley's "Body Politic," p. 209.

[4] Ibid., p. 197. These figures were based on the statistics provided by the Ministry of Health and Welfare in 1960.

[5] See M. J. Meijer, *Marriage and Policy in the Chinese People's Republic* (Hongkong: Hongkong University Press, 1971), p. 61, and appendix 1.

[6] Ibid.; see the appendixes 2, 3, 4, 5 and 6.

[7] Ibid., the appendix 8.

[8] Veronica Pearson, "Population Policy and Eugenics in China" in *British Journal of Psychiatry,* 167 (1995), pp. 1–4.

[9] Ruan Fangfu, "Weile minzu fanrong, yousheng yinggai lifa" (For the sake of national prosperity, eugenics should be legally enacted) in *Xingjiaoyu yu yousheng* (Shanghai: Kexue jishu chubian, 1985), pp. 137–141.

[10] Pearson, "Population Policy and Eugenics in China," p. 2.

[11] Ibid.

[12] Ibid., p. 1.

[13] Tim Beardsley, "China Syndrome: China's eugenics law makes trouble for science and business" in *Scientific American,* March, 1997.

[14] *Xinhua News,* August 11, 1998.

BIBLIOGRAPHY

I. ENGLISH

Adams, Mark B. ed. *The Wellborn Science—Eugenics in Germany, France, Brazil, and Russia.* Oxford: Oxford University Press, 1990.

Appadurai,Arjun. *Modernity at Large.* Minneapolis: University of Minnesota Press, 1996.

Arnove, Robert F. *Philanthropy and Cultural Imperialism: The Foundations at Home and Abroad.* Boston: G.K. Hall & Co., 1980.

Bakhtin, Mikhail. *The Dialogic Imagination*, ed. by Michael Holquist and trans. by Caryl Emerson and Michael Holquist. Austin: University of Texas Press, 1981.

Barlow, Tani E ed. *Formations of Colonial Modernity in East Asia.* Durham: Duke University Press, 1997.

Bartholomew, James R. *The Formation of Science in Japan: Building a Research Tradition.* New Haven: Yale Univ. Press, 1989.

Beardsley, Tim. "China Syndrome: China's Eugenics Law Makes Trouble for Science and Business." *Scientific American*, March, 1997.

Bennett, Adrian Arthur. *John Fryer: The Introduction of Western Science and Technology into Nineteenth Century China.* Cambridge: Harvard University Press, 1967.

Bernstein, Gail L. *Recreating Japanese Women, 1600–1945.* Berkeley: University of California Press, 1991.

Boorman, Howard. *Biographical Dictionary of Republican China* (New York: Columbia University Press, 1967.

Braisted, William Reynolds trans. *Meiroku zasshi: Journal of the Japanese Enlightenment.* Cambridge: Harvard University Press, 1976.

Buck, Peter. *American Science and Modern China, 1876–1936*. Cambridge: Cambridge Univ. Press, 1980.

Budick, Sanford and Wolfgang Iser. *The Translatability of Cultures*. Stanford: Stanford University Press, 1996.

Bullock, Mary B. *An American Transplant: The Rockefeller Foundation and Peking Union Medical College*. Berkeley: University of California Press, 1980.

Chou, Wan-yao, "The Kôminka Movement in Taiwan and Korea: Comparisons and Interpretations" in *The Japanese Wartime Empire, 1931–1945*. ed. by Peter Duus et al. Princeton: Princeton University Press, 1996.

Chow, Rey. "Virtuous Transactions: A Reading of Three Stories by Ling Shuhua" in *Modern Chinese Literature* (vol. 4, 1988), pp. 71–85.

Crawcour, E. Sydney. "Industrialization and Technological Change, 1885–1920" in *The Cambridge History of Japan*. Cambridge: Cambridge Univ. Press, 1988, vol. 6, pp. 385–450.

Croll, Elisabeth. *Feminism and Socialism in China*. London: Routledge and Kegan Paul, 1978.

Cueto, Marcos ed. *Missionaries of Science: the Rockefeller Foundation and Latin America*. Bloomington: Indiana University Press, 1994.

Dikotter, Frank. "Eugenics in Republican China." *Republican China*, 15, no. 1 (Nov. 1989).

———. *Imperfect Conceptions: Medical Knowledge, Birth Defects and Eugenics in China*. New York: Columbia University Press, 1998.

———. *Sex, Culture and Modernity in China*. Honolulu: University of Hawaii Press, 1995.

———. *The Discourse of Race in Modern China*. Stanford: Stanford Univ. Press, 1992.

Dirlik, Arif. "The Ideological Foundations of the New Life Movement: A Study in Counterrevolution." *Journal of Asian Studies*, vol. 34 (1975), no. 4, pp. 945–980.

Dower, John W. *War Without Mercy*. New York: Pantheon Books, 1986.

Duara, Prasenji. "Imperial Nations/National Empires: Ethnography and the Territorial Imperative in East Asia." Unpublished paper,1997.

——— "The Discourse of Civilization and Pan-Asianism," in *Journal of World History*, (2001), vol. 12, no. 1, pp. 99–130.

———. *Rescuing History From the Nation: Questioning Narratives of Modern China*. Chicago: University of Chicago Press, 1995.

Dunn, L. C. A. *Short History of Genetics: The Development of some of the Main Lines of Thought: 1864–1939*. New York: McGraw-Hill, 1965.

Fan Fa-ti. "Nature and National Essence: the Guocui Circle in Early Twentieth-Century China" in *The Moral Authority of Nature*, ed. by Lorraine Daston and Fenando Vidal. Chicago: University of Chicago Press, forthcoming.

Forel, August. *The Sexual Question: A Scientific, Psychological, Hygienic, and Sociological Study*. New York: Medical Arts Agency, 1914.

Galton, Francis. *Hereditary Genius*. London: the Fontana Library, 1962.

Garrett, Shirley S. *Social Reformers in Urban China: The Chinese Y.M.C.A., 1895–1926*. Cambridge: Harvard University Press, 1970.

Gernet, Jacques. *China and the Christian Impact: A Conflict of Cultures*. tran. by Janet Lloyd. New York: the Press Syndicate of the University of Cambridge, 1985.

Gluck, Carol. *Japan's Modern Myths: Ideology in the Late Meiji Period*. Princeton: Princeton University Press, 1985.

Gramsci, Antonio. *Antonio Gramsci—Selections from Cultural Writings*. ed. by David Forgacs and Geoffrey Nowell-Smith. Cambridge: Harvard University Press, 1985.

Haraway, Donna. *Primate Visions: Gender, Race and Nature in the World of Modern Science*. London: Routledge, 1989.

Harootunian, H. D. *Toward Restoration: The Growth of Political Consciousness in Tokugawa Japan*. Berkeley: University of California Press, 1970.

Hasian, Jr., Marouf Arif. *The Rhetoric of Eugenics in Anglo-American Thought*. Athens: the University of Georgia Press, 1996.

Hewa, Soma. *Colonialism, Tropical Disease and Imperial Medicine: Rockefeller Philanthropy in Sri Lanka*. Lanham: University Press of America, 1995.

Howland, D. R. *Borders of Chinese Civilization—Geography and History at Empire's End*. Durham: Duke University Press, 1996.

Hsiung, James and Steven I. Levine. *China's Bitter Vicotory: The War with Japan, 1937–1945*. Armond: An East Gate Book, 1992.

Irie, Akira ed. *The Chinese and Japanese: Essays in Political and Cultural Interactions*. Princeton: Princeton University, 1980.

John Bowers et al. *Science and Medicine in Twentieth-Century China: Research and Education*. Ann Arbor: University of Michigan Press, 1988.

Johnston, William. *The Modern Epidemic: A History of Tuberculosis in Japan*. Cambridge: Harvard University Press, 1995.

Kay, Lily E. *The Molecular Vision of Life:: Caltech, the Rockefeller Foundation, and the Rise of the New Biology*. Oxford: Oxford University Press, 1993.

Kevles, Daniel J. *In the Name of Eugenics—Genetics and the Uses of Human Heredity*. New York: Alfred A. Knopf, Inc., 1985.

Kwok, D. W. Y. *Scientism in Chinese Thought 1900–1950*. New Haven: Yale University Press, 1965.

Lamson, Herbert D. *Social Pathology in China: A Source Book for the Study of Problems of Livelihood, Health and the Family*. Shanghai: The Commercial Press, 1935.

Lindee, M. Susan. *Suffering Made Real: American Science and the Survivors at Hiroshima*. Chicago: University of Chicago Press, 1994.

Liu, Lydia H. *Translingual Practice: Literature, National Culture, and Translated Modernity, China, 1900–1937*. Stanford: Stanford University Press, 1995.

Loftus, Ronald P. "Female Self-Writing: Takamure Itsue's *Hi no Kuni no Onna no Nikki*" *Monumenta Nipponica*, vol. 51 (1996), no. 2, pp. 153–170.

Lu Xun. *Selected Stories.* Peking: Foreign Languages Press, 1972.

Ludmerer, Kenneth M. "American Geneticists and the Eugenics Movement: 1905–1935." *Journal of the History of Biology*, 1969, vol. 2, pp. 337–362.

Ma, Qiusha. *The Rockefeller Foundation and Modern Medical Education in China, 1915–1951.* Dissertation, Case Western Reserve University, 1995.

Masini, Federico. *The Formation of Modern Chinese Lexicon and Its Evolution Toward a National Language:the Period from 1840 to 1898.* Berkeley: University of California, Berkeley, 1993.

Mayr, Ernst. *The Growth of Biological Thought: Diversity, Evolution, and Inheritance.* Cambridge: Harvard University Press, 1982.

McCormack, Gavan and Yoshio Sugimoto, ed. *The Japanese Trajectory: Modernization and Beyond.* Cambridge: Cambridge University Press, 1988.

Meijer, M. J. *Marriage and Policy in the Chinese People's Republic.* Hongkong: Hongkong University Press, 1971.

Najita, Tetsuo and Harry Harootunian, "Japanese Revolt Against the West: Political and Cultural Criticism in the Twentieth Century." *The Cambridge History of Japan*, Vol. 6. Cambridge: Cambridge University Press, 1988.

Najita, Tetsuo and J. Victor Koschmann, eds. *Conflict in Modern Japanese History—the Neglected Tradition.* Princeton: Princeton University Press, 1982.

Nakayama, Shigeru et al. *Science and Society in Modern Japan.* Cambridge, Massachusetts: The MIT Press, 1974.

Nakayama, Shigeru et al. *Science and Society in Modern Japan: Selected Historical Sources.* Cambridge: The MIT Press, 1974.

Nakayama, Shigeru. *Academic and Scientific Traditions in China, Japan, and the West*, trans. by Jerry Dusenbury. Tokyo: Univ. of Tokyo Press, 1984.

Nakayama, Shigeru. *Characteristics of Scientific Development in Japan.* New Delhi: The Center for the Study of Science, Technology and Development, 1977.

Ninkovick, Frank. "The Rockefeller Foundation, China, and Cultural Change." *The Journal of American History*, 1984, 70:4, pp. 799–820.

Ono Kazuko. *Chinese Women in a Century of Revolution, 1850–1950.* Stanford: Stanford University Press, 1989.

Otsubo, Sumiko. "Eugenics in Imperial Japan: Some Ironies of Modernity, 1883–1945" (Dissertation, Ohio State University, 1998)

Pearson, Veronica. "Population Policy and Eugenics in China." *British Journal of Psychiatry*, 167 (1995), pp. 1–4.

Proctor, Robert N. *Racial Hygiene: Medicine under the Nazis.* Cambridge: Harvard University Press, 1988.

Reynolds, David C. *The Advancement of Knowledge and the Enrichment of Life: the Science Society of China and the Understanding of Science in the Early Republic, 1914–1930.* Dissertation, University of Wisconsin-Madison, 1986.

Reynolds, Douglas R. *China, 1898–1912: The Xinzheng Revolution and Japan.* Cambridge: Harvard Univ. Press, 1993.

Richards, Robert J. *Darwin and the Emergence of Evolutionary Theories of Mind and Behavior.* Chicago: University of Chicago Press, 1987.

Risedorph, Kimberly A. "Reformers, Athletes, and Students: the YMCA in China, 1895–1935." Dissertation, Washington University, 1994.

Roberts, Dorothy. *Killing the Black Body: Race, Reproduction, and the Meaning of Liberty*. New York: Pantheon Books, 1997.

Said, Edward W. *Orientalism*. New York: Vintage Books, 1978.

Sakai, Naoki. *Voices of the Past: the Status of Language in Eighteenth-Century Japanese Discourse*. Ithaca: Cornell University Press, 1991.

Sanger, Margaret. *Women and the New Race*. New York: Brentano's, 1920.

Schneider, Laurence A. *Ku Chieh-kang and China's New History: Nationalism and the Quest for Alternative Traditions*. Berkeley: University of California Press, 1971.

Schneider, Laurence. "The Rockefeller Foundation, the China Foundation, and the Development of Modern Science in China." *Social Science and Medicine* 16 (1982): 1217–21.

Soloway, Richard A. *Demography and Degeneration—Eugenics and the Declining Birthrate in Twentieth-Century Britain*. Chapel Hill: Univ. of North Carolina Press, 1990.

Spencer, Herbert. *On Social Evolution: Selected Writings*. ed. by J. D. Peel. Chicago: University of Chicago Press, 1972.

Stepan, Nancy L. *"The Hour of Eugenics" Race, Gender, and Nation in Latin America*. Ithaca: Cornell University Press, 1991.

Su, Ru-chiang "Birth Control in China." Master's Thesis, University of Chicago, 1946.

Sugimoto, Masayoshi and David L. Swain. *Science and Culture in Traditional Japan*. Rutland: Charles E. Tuttle Co., 1989.

Sun, Lung-Kee. "The Presence of the Fin-de-Siècle in the May Fourth Era." *Remapping China: Fissures in Historical Terrain*. Stanford: Stanford University, 1996), pp. 194–209.

Tanaka, Stefan. *Japan's Orient—Rendering Pasts into History*. Berkeley: UC Press, 1993.

Taylor, Griffith. *Environment and Race: A Study of the Evolution, Migration, Settlement and Status of the Races of Man*. London: Oxford University Press, 1927.

Wakabayashi, Bob Tadashi. *Anti-Foreignism and Western Learning in Early-Modern Japan: The New Theses of 1825*. Cambridge: Harvard University Press, 1991.

Wang, Chimin and Wu Lien-Te. *History of Chinese Medicine*. Shanghai: National Quarantine Service, 1936.

Weiner, Michael. *Race and Migration in Imperial Japan: the Limits of Assimilation*. London: Routledge, 1994.

Wright, David. "The Great Desideratum: Chinese Chemical Nomenclature and the Transmission of Western Chemical Concepts." *Chinese Science*, 1997, vol. 14, pp. 35–70.

Yang, Marion. "Birth Control in Peiping," *Chinese Medical Journal*, Vol. 48, no. 8, (August, 1934), pp. 786–791.

Ye, Weili. *Crossing the Cultures: The Experience of Chinese Students in the U.S.A. 1900–1925*. Dissertation, Yale University, 1989.

Yip, Ka-che. *Health and National Reconstruction in Nationalist China: The Development of Modern Health Services, 1928–1937*. Ann Arbor: AAS monograph and occasional paper series, no. 50.

Young, Robert. *White Mythologies: Writing History and the West*. London: Routledge, 1990

Zhongguo kexueshe. *The Science Society of China: Its History, Organization, and Activities*. Shanghai: the Science Press, 1931.

II. CHINESE AND JAPANESE

JOURNALS

Funu Zazhi (1922)
Hua Nian (1932–37)
Kexue (Science) (1915–1945)
Kexue de zhongguo (1933–1937)
Wenhua jianshe (1936–37)
Xin Wenhua (1925–27)
Yousheng Yuekan (1931–1932)
Zhonghua yixue zazhi (the National Medical Journal of China, 1915–1928)
Minzoku eise (1935–1945)
Yûseigaku (1925–1945)
Yûsei undô (1926–1930)

GENERAL MATERIALS: CHINESE

Cai Yuanpei. *Cai Yuanpei quianji*. Beijing: Zhonghua shuju, 1984.

Chen Changheng and Zhou Jianren. *Jinhualun yu shanzhongxue* (Evolution and Eugenics). Shanghai: Commercial Press, 1923.

Chen Dezheng. "Hunyin he shengyu" (Marriage and childbirth). *Funu zazhi*, vol. 8, no. 6, pp. 84–89.

Chen Guofu. Chen Guofu xiansheng qianji. Taipei: Zhengzhong shuju, 1952.

Chen Shengkun. "Zhongguo shengwuxue shiyanpai yu diaocha pai zhi lunzhan" (The debate of Chinese biology: Experiment clique versus Survey clique). *Kexue Yuekan* (Science Monthly), vol. 15, no. 8, 1984, pp. 603–609.

Chen Temao. *Zhongguo zhiwu fenlei xueshi* (History of Chinese botanical taxonomy). Wuchang: Huazhong shifan, 1993.

Ding Wenjiang. "Zhesixue yu pudie" (Eugenics and genealogy). *Gaizao* (Reconstruction), 1919, vol., 3, nos., 4, 5, 6.

Du Yaqian ed. *Dongwuxue dacidian* (Zoology dictionary). Shanghai: the Commercial Press, 1922.

Fan Hongye. "Zong 'gezhi' dao 'kexue'" (From gezhi to kexue). *Ziran bianzhengfa tongxun*, vol. 10, no. 3, 1988, pp. 39–50.

Fang I-chih. *Dongxiqun* (Measuring East and West). Beijing: Zhonghua shuju, 1962.

Fu Weikang. *Zhongguo yixueshi.* Shanghai: Shanghai Zhongyi, 1990.

Gao Mingkai. *Hanyu wailaici cidian* (A dictionary of loan words). Shanghai: Shanghai cishu, 1984.

Hu Dingan. "Guchui yixue geming de huiyi" (Memoir on the campaign for the medical revolution). *Zhuanji wenxue,* vol. 13, no. 5 (1968).

——— *Hu Dingan yishi yanlunji.* Shanghi: Shangwu, 1935.

Huang Fuqing. *Jindai Riben zaihua wenhua ji shehui shiye zhi yanjiu* (The Study of modern Japanese cultural and social enterprises in China). Taipei: Academic Sinica, 1982.

Kang Youwei. *Riben shumu zhi.* (Kang Nanhai xiansheng yizhu huikan, vol. 11). Taipei reprint.

Kexue chubanshe ed. *Zhongguo Xiandai kexuejia zhuanji* (Biography of modern Chinese Scientists). Beijing: Kexue chuban, 1991.

Liang Qichao. *Kang Nanhai zhuan* (Biography of Kang Youwei). Shanghai: Guangzhih shuju, 1906.

Liang Qichao. *Yinbing shi heji.* Shanghai: Zhonghua shuju, 1932.

Liu Zhaomin. *Zhonghua shengwu xueshi* (The history of Chinese biology). Taipei: Shangwu, 1991.

Long Guanghai. *Zhongguo renkou.* Taipei: Zhonghua wenhua, 1955.

Lu Yudao et al. *Kexue de minzu fuxing* (Scientific national revival). Shanghai: Zhongguo kexueshe, 1937.

Lu Yudao et al. *Zhongguo zhanshi xueshu.* Shanghai: Zhengzhong shuju, 1946.

Ma Xiangbo. *Yiri yitan.* Shanghai: Shangwu, 1936.

Ma Zuyi. *Zhongguo fanyi jianshi: wusi yiquian bufen.* Beijing: Zhongguo duiwai fanyi chuban gongsi, 1984.

Pan Guangdan. *Pan Guangdan wenji* (Pan's collected works). Beijing: Beijing daxue, 1993.

———. *Pan Guangdan minzu yanjiu wenji.* Beijing: Minzu chubanshe, 1995.

———. *Yousheng yu kangzhan* (Eugenics and the anti-Japanese war). Chongqing: Shangwu shuju, 1944.

———. *Yousheng yu kangzhan* (Eugenics and the anti-Japanese war). Chongqing: Shangwu shuju, 1944.

Qian Yiji. *Beizhuanji* (The collection of inscriptions and biographies). Taipei reprint, year unknown.

Quan Hansheng. "Qingmo di 'xixue yuanchu zhongguo' shuo" (The theory of Chinese origin of Western Learning in the late Qing). *Lingnan xuebao,* (Guanzhou, 1935), vol. IV, no. 2, pp. 57–102.

Ren Hongjun. "Shinianlai Zhongjihui shiye de huigu"(Rreview of the China Foundation's ten-year achievements). *Dongfang zazhi.* vol. 32, no. 7, 1935, pp. 19–25.

Ren Zhuoxuan. "Ping Youshengxue yu huanjinglun de lunzheng." *Ershi shiji.* vol. 1, (1931), no. 1, pp. 57–124.

Ruan Fangfu. "Weile minzu fanrong, yousheng yinggai lifa" (For the sake of national prosperity, eugenics should be legally enacted) in *Xingjiaoyu yu yousheng*. Shanghai: Kexue jishu chubian, 1985, pp. 137–141.

Ruan Yuan. *Chouren zhuan*. Taipei: Shangwu, 1982.

Sanetô Keishû. *Zhongguoren liuxue ribenshi* (The history of Chinese studied in Japan), translated by Tam Ruqian and Lin Qiyan. HongKong: The Hongkong Chinese University Press, 1982.

Se Lu. "Chaner zhixian yu zhongguo" (Birth control and China). *Funu zazhi*. vol. 8, no. 6, pp. 10–14.

Shen Zonghan and Liu Tuo ed. *Kexuehua yundong kwangbo yanjiangji*. Peiping: Zhongguo kexuehua yundong xiehui peiping fenhui, 1936.

Sun Benwen et al. *Zhongguo zhanshi xueshu*. Shanghai: Zhengzhong shuju, 1946.

———. "Zailun wenhua yu youshengxue." *Shehui xuekan*. vol. 1 (1929), no. 2.

Sun Yat-sen. *Sanming zhuyi*. Taipei: Zhengzhong shuju, 1978.

Tan Ruqian. *Zhongguo yi riben shu zonghe mulu*. Hongkong: Zhongwen daxue, 1980.

Wang Hui. "'Sai xiansheng' zai zhongguo de mingyun: zhongguo jinxiandai sixiangzhong de 'kexue' gainian jiqi shiyong (The fate of 'Mr. Science' in China: the concept of 'kexue' and its application in Modern Chinese thought)." *Xueren*. vol. 1, 1991, pp. 49–123.

Wang Jingwe., Wang Jingwei xiansheng heping yundong yanlunji. Guangzhou: Zhongshan ribao, 1939.

Wang Rongbao and Ye Lan ed. *Xin Erya*. Shanghai: Wenming shuju, 1906.

Wang Yangzong. "Qingmo yizhi shuhui tongyi shuyu gongzuo shuping (A comment on the work of the late Qing Educational Association of China in standardizing technical terminology). *Zhongguo keji shiliao*. 1991, vol. 12, no. 2, pp. 9–19.

Wu Shenyuan. *Zhongguo renko sixiang shigao* (A history of Chinese population thought). Zongqing: Xinhua shudian, 1986.

Xiong Bingzhen. "Zhongguo jinshi de xinshenger zhaohu" in *Zhongguo jinshi shehui wenhuashi lunwenji*. Taipei: Institute of Modern Chinese History of Academica Sinica, 1992.

Xu Yongping. *Chen Guofu Zhuan* (The Biography of Chen Guofu). Taipei: Zhengzhong shuju, 1978.

Xu Zhenhua. "Youshengxue yu zhongguo" (Eugenics and China) in *Tung-chi Acta Medica* (Tongji yixue quarterly), vol. 1 (1931), nos. 1 and 2, pp. 16–93.

Yan Renying ed. *Yang Chongrui Boshi* (Doctor Yang Chongrui). Beijing: Zhongguo Xiehe yike daxue, 1990.

Yu Fenggao. *Lu Xun zawen yu kexushi* (Lu Xun's essays and history of science). Hangzhou: Zhejian wenyi, 1986.

Zhang Jingyuan. *Zhongguo renkou wenti*. Taipei: Zhongguo renkou xuehui, 1959.

Zhang Junjun. *Minzu suzhi zhi gaizao*. Chongqing: the Commercial Press, 1944.

Zhang Ruogu ed. *Ma Xiangbo xiansheng nianpu*. Taipei: Wenhai reprint.

Zhang Xichen ed. *Xin xingdaode taolunji.* Shanghai: Kaiming, 1925.
Zhen Fu. "Lu Xun he Zhang Xichen." *Dushu zazhi,* vol.1 (1979), no. 1, pp. 87–88.
Zhongguo kexuehua yundong xiehui. *Kexue yu zhongguo* (Science and China). Shanghai: Zhongguo kexuehua yundong xiehui, 1936.

General materials: Japanese

Ban Tadayasu ed. *Tekijuku to Nagayo Sensai.* Tokyo: Sôgensha, 1987.
Fujino Yutaka. *Nihon fashizumu to iryô: hansenbyô o meguru jisshôteki kenkyû* (Japanese fascism and medical care: an evidential investigation in connection with leprosy). Tokyo: Iwanami shoten, 1993..
Fukuzawa Yukichi. *Fukuzawa Yukichi shû.* Tokyo: Chikuma shobô, 1977.
Furuhata Tanemoto. "Ketsuekigata yori mita nihonjin (The Japanese seen from the blood types)." *Nihon minzoku,* ed. by the Tokyo Anthropological Association. Tokyo: Iwanami shoten, 1935, pp. 85–101.
Furuya Yoshio. *Kokudo, jinkô, ketsueki.* Tokyo: Asahi shimbunsha, 1941.
Furuya Yoshio. *Kokudo, jinkô, ketsueki.* Tokyo: Asahi shimbunsha, 1941.
Gotô Shimpei. "Kokka eisei genri." *Kagaku no shisô,* II. Tokyo: Chimaku shobô, 1964, pp. 289–303.
Hiroshige Tetsu. *Kagaku no shakaishi: Kindai nihon no kagaku taisei.* Tokyo: Chûô kôron, 1973.
Horiuchi Gôji, *Kindai kagaku shisô no keifu* (The Genealogy of modern scientific thought). Tokyo: Shibundô, 1964.
Hosaka Yûichirô ed. *Dôjinkai yonjûnen shi* (The Forty years of the Benevolence Association). Tokyo: Dôjinkai, 1943.
Ikeda Shigenori. *Shidôsha minzoku no yuseigaku teki ishin.* Tokyo: Nihon shuppansha, 1942.
Imai Seiichi ed. *Jûgonen sensô to higashiajia.* Vol. 2 of *Taikei: Nihon gendaishi.* Tokyo: Nihon hyôronsha, 1979.
Inoue Takeshi. "Kagaku no taisei to kagakusha no shisô" *Kagaku no shisô.* Vol. 25 of *Gendai nihon shisô taike.* Tokyo: Chimaku shobô, 1964), pp. 7–54.
Ishihara Jun. "Shakai jijô to kagaku teki seishin" in *Kagaku no shisô.* Vol. 25 of *Gendai nihon shisô taike,* pp. 121–131.
Ishii Tomoyuki and Ishihara Tasuro, *Seibutsugaku* (Tokyo: Sanryû shobô, 1935.
Ishijima Noriyuki. "Kô-Nichi sensô." Vol. 2 of *Taikei: Nihon gendaishi.* Tokyo: Nihon hyôronsha, 1979.
Katô Shizue. *Aru josei seijika no hansei.* Tokyo: PHP Kenkyûjo, 1981.
Katô Shûichi. "Meiji shoki no hon'yaku." *Hon'yaku no shisô* (*Nihon kindai shisô taikei,* 15. Tokyo: Iwanami Shoten, 1991), pp. 342–380.
Katô Shûichi. "Meiji shoki no buntai." *Buntai* (*Nihon kindai shisô taikei,* 16). Tokyo: Iwanami Shoten, 1991, pp. 449–481.
Masaaki Sogo and Yoshifumi Hida, *Meiji no kotoba jiten.* Tokyo: Tokyodô, 1986.
Morioka Kenji. *Kindaigo no seiritsu: Meiji jiki goihen.* Tokyo: Meiji shoin, 1969.

Nagai Hisomu. "Ketsu to bunka to nihon minzoku." *Inochi*, (October, 1937), pp. 202–221.
Nagai Hisomu. "Pekin daigakusei ni tsugu" was published in *Chûô kôron*, no. 654 (Feb., 1942), pp. 54–66.
Nagai Hisomu. "Sabetsukan to byôdôkan." *Tôa renmei* (East Asian League), vol. 2, no. 2 (Feb., 1940), pp. 90–98.
Nagai Hisomu. *Dô to shizen* (Way and Nature). Tokyo: Jinbun shoin, 1936.
Nagai Hisomu. *Seibutsugaku to tetsugaku to no sakai*. Tokyo: Bunka seikatsu kenkyûkai, 1926.
Nagai Hisomu. *Seimeiron*. Tokyo: Rakuyôdô, 1915.
Nagai Hisomu. *Shizenkan yori jinseikan e* (From a view of nature to a view of life). Tokyo: Jinbun shoin, 1933.
Nagai Hisomu. *Yûseigaku gairon*. Tokyo: Yûsankaku, 1936.
Naigai Hisomu. *Dô to shizen*. Tokyo: Jinbun shoin, 1936.
Narita Ryûichi. "Shintai to kôshû eisei." *Kôza sekaishi*. Rekishi gaku kenkyu kai ed., vol.4. Tokyo:Tokyo University Press, 1995, pp.375–401.
Nihon joi shikai. *Nihon joi shi* (The history of Japanese female physicians). Tokyo: Chûo kôron, 1962.
Nishi Amane. *Nishi Amane zenshû*, Tokyo: Munetaka shobô, 1960.
Oguma Eiji. *Tan'itsu minzoku shinwa no kigen* (the Myth of the Homogeneous Nation). Tokyo: Shinyôsha, 1996.
Ogura Kinnosuke. *Kagaku ron: Sûgakusha no kaisô*. Tokyo: Keisô shobô, 1974.
Ono Kazuko. "Shimoda Utako to Hattori." *Kindai Nihon to Chûgoku*, ed. Takeuchi Yoshimi and Hashikawa Bunzô. Tokyo: Asahi shimbunsha, 1974, vol. I, pp. 201–221.
Ôta Tenrei. *Nihon sanji chôsetsu hyakunen shi* (The hundred-year history birth control in Japan. Tokyo: Yamahan kagaku, 1976.
Saitô Shôji. *Ninhon shakaigaku seiritsu shi no kenkyû*. Tokyo: Fukumura, 1976.
Sanetô Keishû. *Riben wenhua ge zhongguo de yingxiang* (the Influence of Japanese Culture in China, and the original title *Nihon bunka no Shina e no eikyô*, Tokyo, 1940). translated by Zhang Mingsan. Shanghai: Xinshenbao guan, 1944.
Shimizu Katsuyoshi. *Nihon kôshû eiseishi* (The History of public health, the early half of the Shôwa period). Tokyo: Fuji shuppan, 1989.
Shirayanagi Shûko. *Nihon minzoku ron*. Tokyo: Chikura shobô, 1942.
Shôji,Yamada. "Jiyû minkenki ni okeru kô-a-ron to datsu-a-ron: Ajia shugi no keisei o megutte." *Chôsenshi kenkyû kenkyûkai ronbunshû*, vol. 6, 1969, pp. 40–63.
Sugimoto Tsutomu. *Nihon Honyakugo shi no kenkyû*. Tokyo: Hachisaka Shobô, 1983.
Suzuki Naoko ed. *Sengo boseino yukue*. Tokyo: Domesu Shuppan, 1985.
Suzuki Naoko ed., *Sengo bosei no yukue*. Tokyo: Domesu Shuppan, 1985.
Suzuki Zenji, *Nihon no yûseigaku: so no sishô to undô no kiseki* (Japanese eugenics: its ideas and locus of movement). Tokyo: Sankyô, 1983.

Suzuki Zenji. "Geneticists and the Eugenics Movement in Japan." *Japanese Studies in the History of Science*, (1975) vol. 14.

Tanabe Hajime. *Tanabe Hajime zenshu.* Tokyo: Chikuma shobô, 1963.

The Japanese Konoe Cabinet. *Kokka sôdôin* (National mobilization). Tokyo: Kokufûkai, 1937.

The Kôseishô. *Kôseishô nijûnen shi.* Tokyo: the Kôseishô,1960.

Yamamoto Senji. *Yamamoto Senji zenshû.* Tokyo: Chobunsha, 1979.

Yano Fumio. *Nihon buntai moji shinron.* Tokyo: Hôchisha, 1886.

Yoshimasu Shûfu. *Yûseigaku* (Eugenics). Tokyo: Nankôdô, 1961.

Yoshimasu Yasuo et al., *Yûseigaku* (Eugenics) (Tokyo: Nankodo, 1961

Yoshioka Yayoi. *Kono jû nen kan* (At the time of this decade). Tokyo: Gakufû shoin, 1952.

Yoshioka Yayoi. *Yoshioka Yayoi den.* Tokyo: Nihon Tosho Center, 1998.

Yoshioka. "Genzai no nihon ni wa dai kinmotsu." *Fujin sekai*, vol: 17, no. 4, (April, 1922), pp. 103–104.

Yoshioka. *Fujin ni atau.* Tokyo: Ôzorasha, 1997, reprint. Originally published by Hôdô shuppansha, 1943.

Index

For Product Safety Concerns and Information please contact our EU
representative GPSR@taylorandfrancis.com
Taylor & Francis Verlag GmbH, Kaufingerstraße 24, 80331 München, Germany

www.ingramcontent.com/pod-product-compliance
Ingram Content Group UK Ltd.
Pitfield, Milton Keynes, MK11 3LW, UK
UKHW040102200926
475702UK00001B/4

* 9 7 8 0 4 1 5 9 3 3 6 6 7 *